BORON REINFORCED ALUMINUM SYSTEMS. PART II

VOLUME ELEVEN
MATERIALS TECHNOLOGY SERIES

Edited by

CARLOS J. HILADO

a TECHNOMIC publication
TECHNOMIC Publishing Co., Inc.

BORON REINFORCED ALUMINUM SYSTEMS. PART II

VOLUME ELEVEN

MATERIALS TECHNOLOGY SERIES

a TECHNOMIC® publication

PO Box 913, 265 Post Road West, Westport, CT 06881

Printed in U.S.A.
Library of Congress Card No. 74-83231
Standard Book No. 87762-312-0

To Linda

CONTENTS

PUBLISHER'S NOTE

The Effect of Ply Lay-Up Sequence on the Fracture Toughness of Boron Aluminum

KARL M. PREWO

United Technologies Research Center
East Hartford, Connecticut 06108

(Received September 3, 1977)

ABSTRACT

The fracture toughness of 0/90° cross ply boron reinforced aluminum composites has been investigated using prenotched compact tension and center notched tensile specimens. The effects on fracture toughness of specimen geometry, the relative numbers of 0° and 90° plys, and the lay-up sequence, have been measured. Ply lay-up sequence has been shown to affect fracture morphology and in turn alter the value of toughness. This has been related to the degree of matrix deformation which can take place to blunt the advancing crack. The relative numbers of 0° and 90° plys has been shown to be the major controlling factor and simple calculations based on a rule of mixtures approach provide reasonable agreement with the experimental data.

INTRODUCTION

In a recent report [1], it was shown that the fracture toughness of boron reinforced aluminum composites is quite high, when compared with monolithic engineering materials. It was also shown that actual values of composite fracture toughness were very orientation and ply lay-up sequence dependent. The herein reported investigation examined the ply lay-up sequence dependence in greater detail and also compared data obtained from two entirely different test specimens.

EXPERIMENTAL PROCEDURE

All composite specimens were fabricated from plasma sprayed tapes by the "Quick Bond" process using the conditions of 550°C, 68.9 MPa, 10 min in air. The fiber was 140μ diameter boron and the matrix 6061 aluminum. Fiber content was 50–55%.

Two different types of fracture toughness specimens were tested. Compact tension (CT) type specimens were fabricated to ASTM specifications, except for the thickness which in this case was about 0.381 cm. These specimens, Figure 1a, were

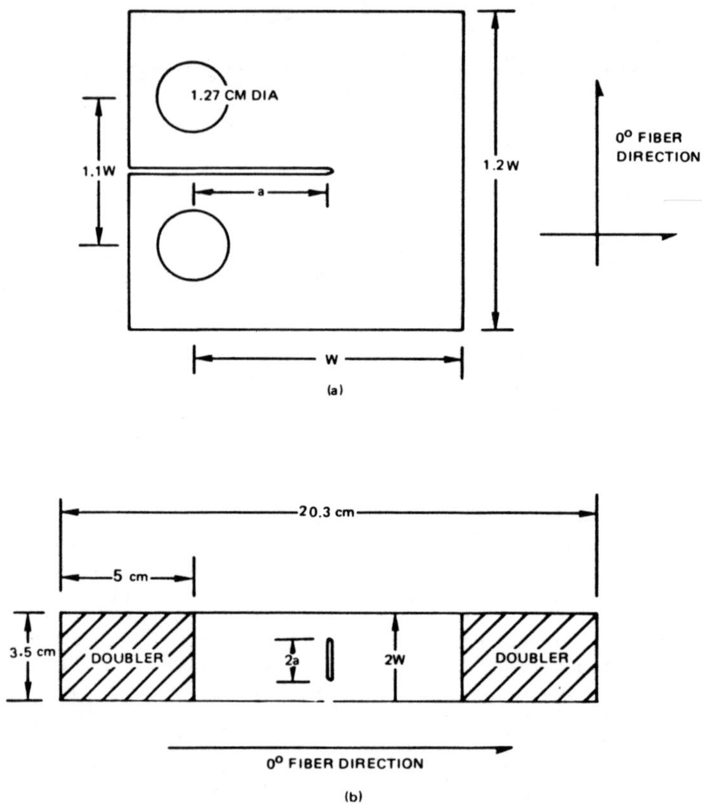

Figure 1. *Pre-notched test specimen.*
(a) Compact tension, $W = 5.08$ cm, thickness = 0.38 cm
(b) Center notched, thickness = 0.2 cm

similar to those used previously [1] except for the present closer conformance to ASTM convention. The second specimen type was a center notched (CN) configuration, Figure 1b. In this case, because of the much higher loads required to fracture this type of specimen, and gripping limitations, specimen thickness was limited to 0.196 cm.

In both specimen types, the notches were introduced by EDM technique. The resultant notch tip radius was 150μ in all cases. Additional CN specimens were fabricated with circular holes to compare the effects of hole geometry.

Specimen ply lay-up sequence was varied in the CT specimens and held constant for the CN specimens. The four sequences used in this study, along with their abbreviated notations are given in Table 1.

Table 1. Specimen Lay-Up Sequences.

Type	Abbreviated Notation	Detail Layup of One-Half of Specimen Thickness*
A	$[(90/0)_6]_s$	90/0/90/0/90/0/90/0/90/0/90/0 – Center
B	$[(90/0_2)_4]_s$	90/0/0/90/0/0/90/0/0/90/0/0 – Center
C	$[(90/0_3)_3]_s$	90/0/0/0/90/0/0/0/90/0/0/0 – Center
D	$[90_6 0_6]_s$	90/90/90/90/90/90/0/0/0/0/0/0 – Center

*Layup sequence is symmetrical about the center of the specimen

The mechanical test procedure for the CT specimens included the use of an extensometer attached across the edges of the notch on the left side of the specimen as shown in Figure 1. Load was applied at a constant rate and recordings of applied load versus extensometer displacement were made. This displacement will be referred to as crack opening displacement (COD). In addition visual and auditory monitoring of each specimen during test permitted notation of the loads at which specimen acoustic emission and visible crack growth occurred.

All calculations of stress intensity (K) values were made based on applied loads and specimen geometry as reported in [1] using the data available in [2, 3].

Parallel sided unnotched tensile specimens were also fabricated and tested.

EXPERIMENTAL RESULTS

Compact Tension Specimens

The load versus COD recorded traces for three of the specimens tested in this program are presented in Figure 2. In all cases specimen audible acoustic emission began at very low load levels and well in advance of any visible crack growth. In the case of specimen set 2789 and two out of three specimens in set 2790, visible crack extension occurred prior to the maximum load experienced in the test. For the remaining specimens, crack growth occurred at the maximum load attained during the test. Values of stress intensity associated with the loading history of each specimen are tabulated in Table 2.

The direction of crack growth did not always coincide with the original machined notch. The two types of observed behavior are shown in Figure 3 and crack growth directions are recorded in Table 2. In the cases of all specimens where 50% of the plys were at 90° (Types A and D), crack growth for all values of (a/w) was colinear with the original notch, i.e. in the 90° direction. For those specimens where only 33% of the plys are oriented at 90° (Type B), only the longer notches

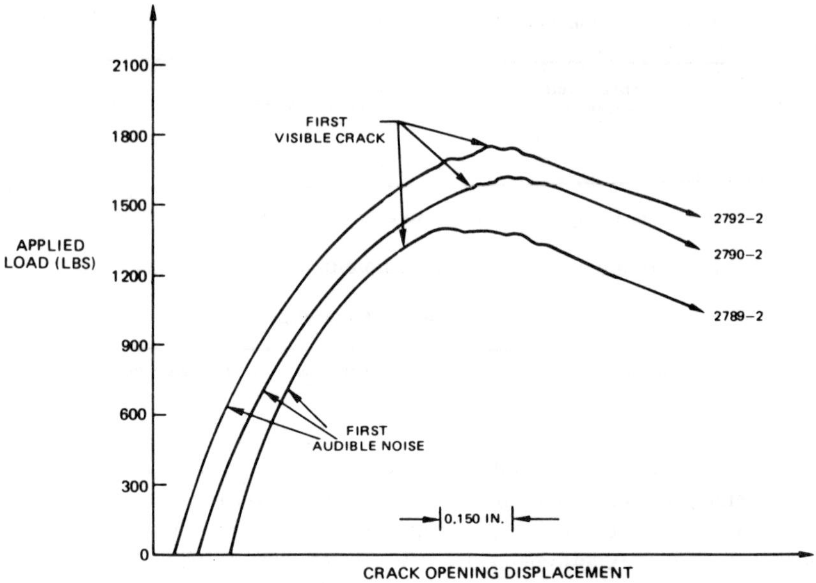

***Figure 2.** Applied load vs crack opening displacement for compact tension specimens.*

***Table 2.** Composite Tensile Specimen Data.*

Specimen Number	Orientation	(a/w)	K_{noise} MPa$\sqrt{m}$	Stress Intensity* $K_{visible\ crack}$ MPa$\sqrt{m}$	$K_{maximum}$ MPa$\sqrt{m}$	Direction of Crack Growth
2789-1	$[(90/0)_6]_s$	.304	27.2	63.2	63.7	90
-2	Type A	.456	25.5	56.1	58.7	90
-4		.597	28.6	54.6	56.7	90
2790-1	$[(90/0_2)_4]_s$	.336	22.5	64.2	64.2	0
-2	Type B	.457	52.3	69.5	71.5	90
-4		.608	20.9	67.0	70.5	90
2791-1	$[(90/0_3)_3]_s$	.304	25.2	57.9	57.9	0
-2	Type C	.454	14.1	64.0	64.2	0
-4		.605	34.5	73.4	73.3	0
2792-1	$[90_6 0_6]_s$	.314	30.5	77.4	77.4	90
-2	Type D	.447	19.5	73.7	73.7	90
-4		.633	28.5	66.0	66.0	90

* 1 ksi$\sqrt{in.}$ = 1.099 MPa$\sqrt{m}$

permitted colinear crack growth and for those specimens where 25% of the plys were at 90° (Type C), all crack extension occurred in the 0° direction. It is thus shown that the ability to divert crack extension from the original notch plane is

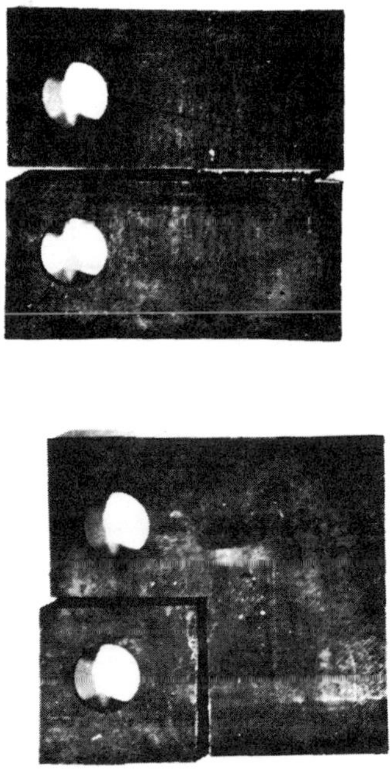

Figure 3. Fractured compact tension specimens.

(a) Spec 2792-2, crack growth colinear with notch (90°)

(b) Spec 2790-1, crack growth perpendicular to notch (0°)

sensitive to both ply lay-up and specimen geometry.

The effect of specimen ply lay-up sequence on material fracture toughness was found to be quite large. The K values obtained for specimen sets 2789 and 2792 are compared in Figure 4. The former specimens consist of alternate 0° and 90° plys throughout (Type A), while the latter consist of an inner core of 0° plys with the 90° plys segregated to the outer surfaces (Type D). Both types of composites exhibited audible acoustic emission at a stress intensity level of about 27.5 MPa$\sqrt{m}$, however, the stress intensity values for crack growth were significantly higher for set 2792. In addition, crack growth did not occur until the maximum applied load had been achieved for these specimens, while for the alternating ply specimen set 2789, crack growth occurred prior to maximum load.

The values of stress intensity applied to CT specimens at the point of crack extension are plotted in Figure 5 as a function of (a/w). In the case of those specimens for which crack growth is parallel to the initial notch (90°), the K value decreases with increasing a/w while for those specimens in which crack growth is in the 0° direction, effectively blunting the original notch, the K value increases with increasing a/w.

Examination of specimen fracture surfaces by scanning electron microscopy revealed substantial differences in fracture plane morphology. The most apparent contrast is between specimen sets 2789 (Figure 6) and 2792 (Figure 7). In the former Type A case, there is very little fiber pull out on the fracture surface; crack growth was confined to the plane of the initial EDM notch. In contrast, for the completely segregated ply lay-up Type D specimens, Figure 7, a considerable amount of fiber pull out is observed. The grouping together of the 0° plys reduced the constraining effect of the 90° plys in confining fracture to a single plane. Specimen 2790-2, Figure 8, also exhibited some fiber pull out due to a reduction in number of the 90° plys to only 33% of the total lay-up, Type B.

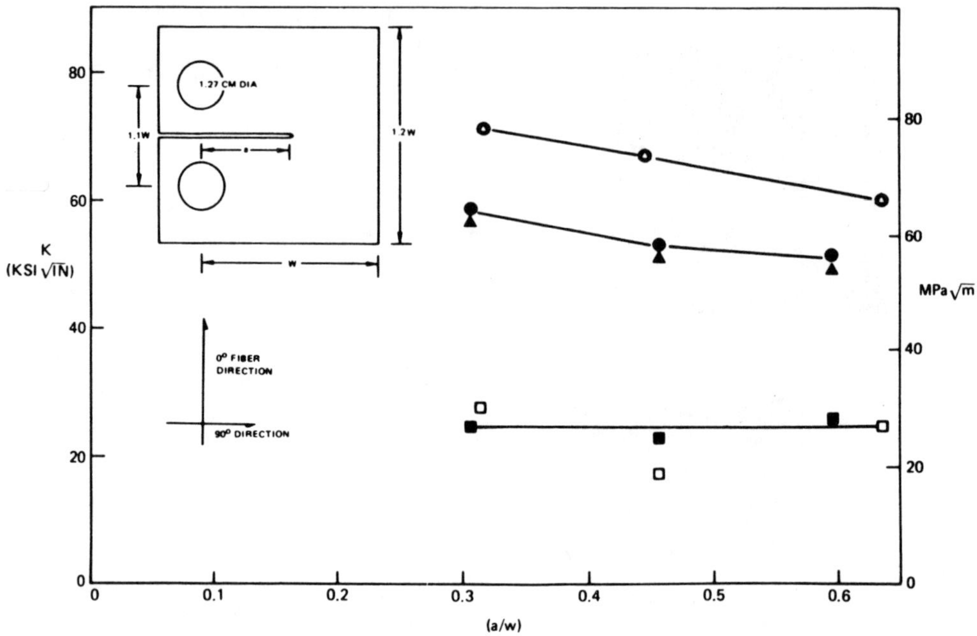

Figure 4. Stress intensity (K) as a function of (a/w) for CT specimens.

Composite	K_{Noise}	$K_{Visible\ Crack}$	K_{max}	*Lay-Up*
2789	■	▲	●	$[(90/0)_6]_S$
2792	□	△	○	$[90_6 0_6]_S$

Center Notched Specimens

The center notched specimens were not tested with an extensometer in place, so only applied load vs machine crosshead displacement curves were recorded. These were linear to the point of fracture for all of the specimens tested and fracture occurred in a sudden manner across the total specimen width. The fracture appearance of a specimen with a 0.66 cm long slit and of a specimen with a 0.64 cm diameter hole are shown in Figure 9. In both cases specimen fracture propagated on a plane normal to the tensile axis with only a small amount of fiber pull out occurring. This fracture morphology agrees well with the CT specimens which had the same ply lay-up sequence, i.e. Type A.

The mechanical test data are presented in Table 3. The tensile strength of this material in the unnotched condition is probably greater than the numbers shown in the table. This is due to the failure of these specimens under their doubler pads, which was caused by excessive gripping pressure during test. The fracture of

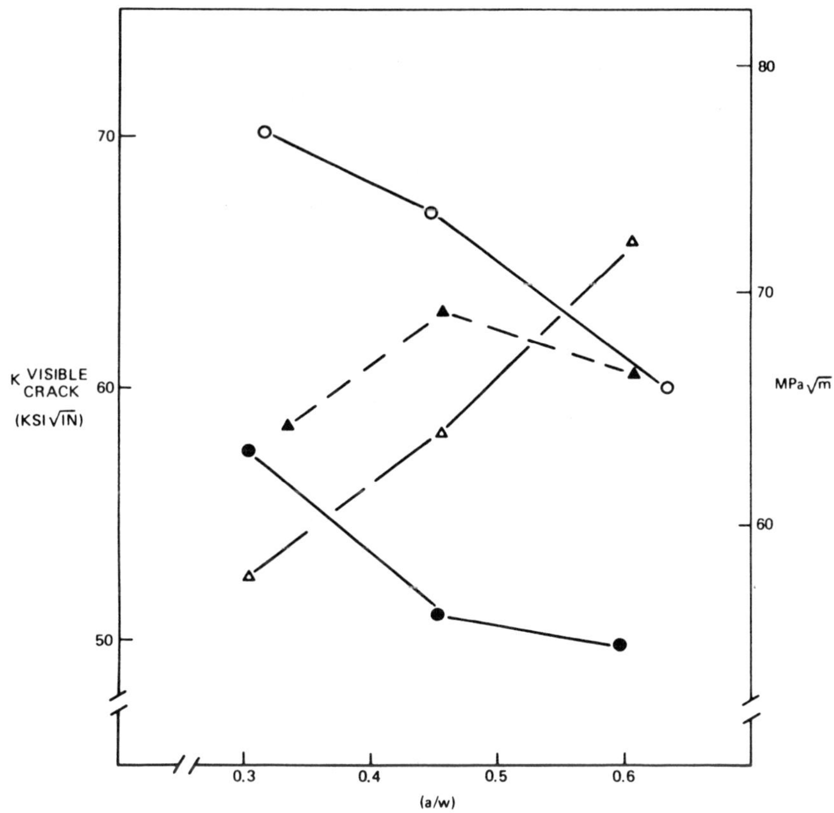

Figure 5. Stress intensity applied for crack growth as a function of a/w.

Composite	*Lay-Up*		*Crack Growth Direction*
2789	$[(90/0)_6]_S$	●	*90°*
2790	$[(90/0_2)_4]_S$	▲	*0° or 90°*
2791	$[(90/0_3)_3]_S$	△	*0°*
2792	$[90_6 0_6]_S$	○	*90°*

specimen 2793-4b, which had a short notch with $a/w = 0.09$, also occurred under the doublers, however all other notched specimens failed at the notch or at the point of minimum remaining cross sectional area, in the case of the circular holes. Composite net ultimate tensile strength was calculated for each specimen based on the specimen cross section minus the notch. The data in Table 3 indicate that this remaining net strength is substantially greater for those specimens with circular holes as compared to those with slits.

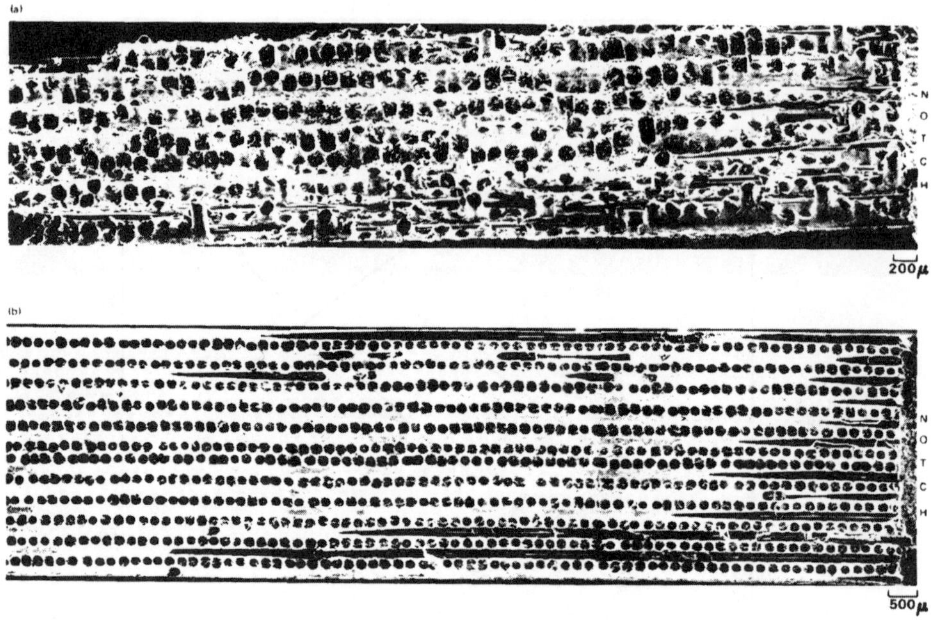

Figure 6. Fracture surface of specimen 2789-2.
SEM angle = 50° (a) and 0° (b)
Ply lay-up sequence $[(90/0)_6]_s$

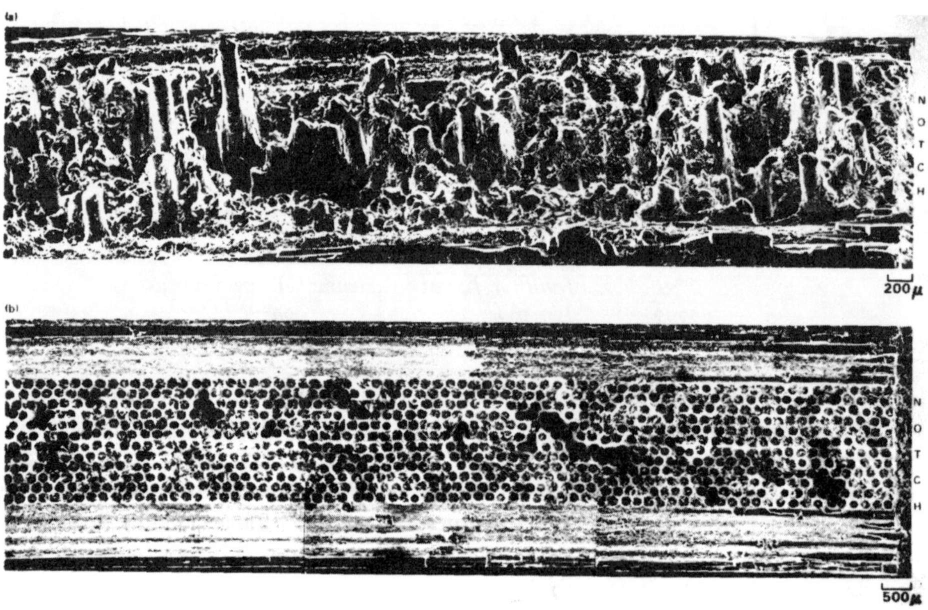

Figure 7. Fracture surface of specimen 2792-2.
SEM angle = 60° (a) and 0° (b)
Ply lay-up sequence $[90_6 0_6]_s$

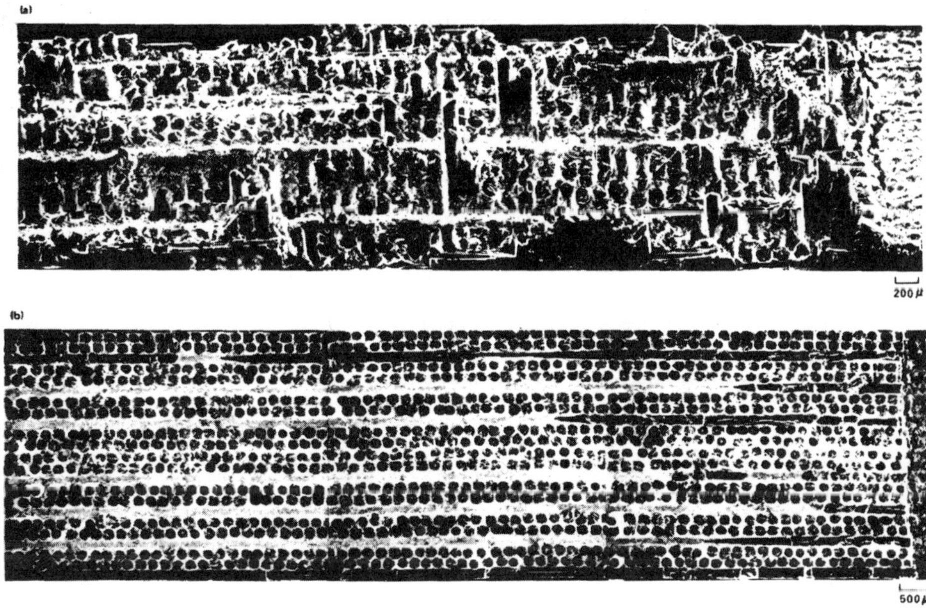

Figure 8. *Fracture surface of specimen 2790-2.*
SEM angle = 50° (a) and 0° (b)
Ply lay-up sequence $[(90/0_2)_4]_s$

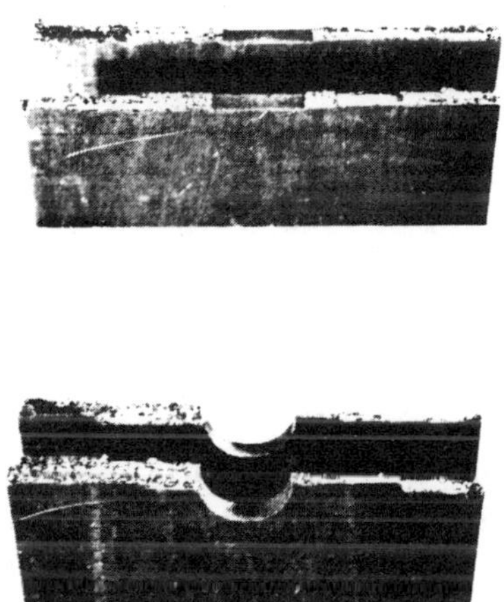

Figure 9. *Fractured center notched specimens.*
(a) Spec 2793-1a, total crack length = 0.66 cm
(b) Spec 2793-1b, hole diameter = 0.64 cm

Stress intensity values were calculated for each of the specimens which failed at a slit and these are compared with the fracture toughness values obtained for CT specimens of the same ply lay-up in Figure 10. The CN specimens fractured at considerably lower levels of stress intensity than the CT specimens.

Unnotched Tensile Specimens

The tensile data obtained from unnotched specimens are shown in Table 4. A comparison was made between two lay-up sequences, similar to Types A and D described in Table 1. The strengths obtained were found to be essentially independent of lay-up and the only significant difference between the two sets of composite tensile data was found to be in the values of strain to failure. This difference, however, was simply due to the fact

Table 3. Center Notched Specimen Data.

Specimen	Notch	(a/w)	Net UTS MPa	$K_{maximum}$ MPa$\sqrt{m}$
2793-4a	No		529.8	-
2793-3a	No		751.0	-
2793-1a	0.66 cm Slit	0.19	475.4	40.1
2793-1b	0.64 cm Diam. Hole	0.18	573.2	-
2793-2a	1.32 cm Slit	0.38	442.3	44.0
2793-2b	1.28ccm Diam. Hole	0.37	571.9	-
2793-3b	1.05 cm Slit	0.3	509.9	49.5
2793-4b*	0.32 cm Slit	0.09	519.5	-

*This specimen fractured at a section removed from the notch

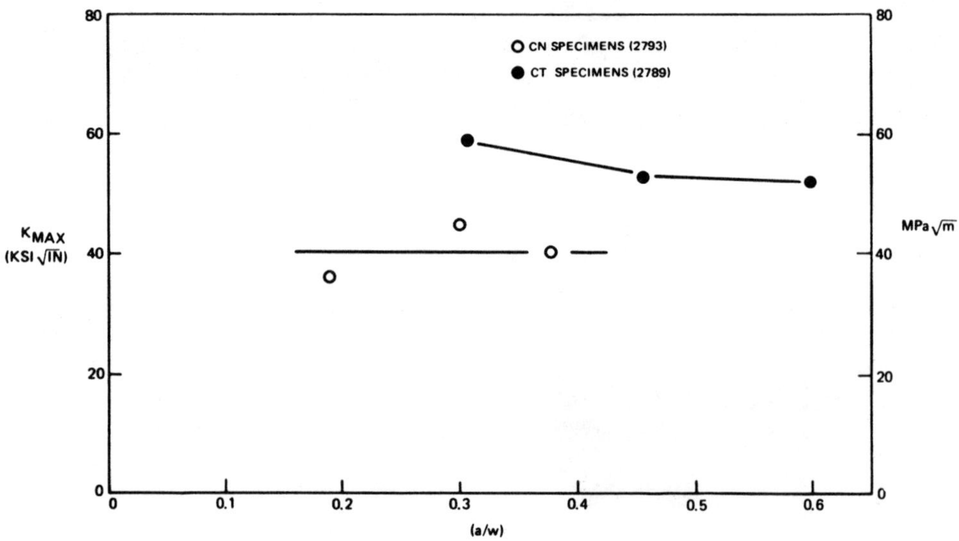

***Figure 10.** Fracture toughness comparison of CN and CT specimens with alternating 0/90 ply lay-up. Crack growth direction is 90°*

that for one set of specimens the lowest strength values happened to occur for specimens with strain gages, while for the other set the opposite held true and the higher strengths occurred for gaged specimens. Additional examination of the shapes of the tensile stress strain curves failed to find any other distinguishing features so that it is most reasonable to conclude that tensile performance was not dependent on lay-up sequence.

Table 4. Unnotched Tensile Specimen Data.

Lay-Up Sequence		UTS		E		ϵ_f
		10^3 psi	MPa	10^6 psi	GPa	%
$[90_3/0_3]_s$		102	700	23.7	163	0.65
Similar to C 2792		101	695	23.0	159	0.60
Type D		106	732	23.6	163	0.65
		108	744			
		113	781			
		124	854			
	Mean Value	109	751			
$[(90/0)_3]_s$		143	983	24.7	170	0.81
Similar to C 2789		122	843	24.2	167	0.75
Type A		115	794	23.8	164	0.68
		111	765			
		115	795			
		105	724			
	Mean Value	118	817			

SUMMARY AND CONCLUSIONS

It has been demonstrated that the fracture toughness and mode of crack growth for cross ply boron reinforced aluminum specimens is dependent on the relative percentages of 0° and 90° plys. This was expected and is consistent with previous findings for B/Al composites having other orientations [4]. The percentage of 0° plys in the composite is the most important factor and, as shown in Figure 11, a simple rule of mixtures approach and a 0° ply controlled failure calculation bound the data obtained in this report. For this representation the fracture toughness values for all 0° and all 90° ply composites were taken from prior studies [1, 5] conducted with material fabricated at United Technologies Research Center using the same procedure and starting materials as those used for the herein described report. Although included in the data plot, the stress intensity value obtained at fracture of the C type composite lay-up is conservative since fracture did not propagate in a direction parallel to the initial machined notch, but instead at right angles to it.

The tendency for crack growth to occur at right angles to the premachined notch (referred to as 0° crack growth in this report) varied with both the percentage of 0° plys and the value of (a/w) for the CT specimen type. This type of behavior has already been analyzed [1] using a simple beam calculation. As in the previous case, a value of composite transverse strength which is independent of (a/w) can be obtained for all three specimens of type C lay-up. It is this strength which must be exceeded to cause the composite to fracture by crack growth parallel to the 0° direction. The values of transverse composite strength obtained for Type C specimens with (a/w) equal to 0.30, 0.45 and 0.60 are 225 MPa, 256 MPa and 242 MPa respectively, for an average value of 242 MPa. This value is lower than the 350 MPa value of strength expected for the transverse strength of this

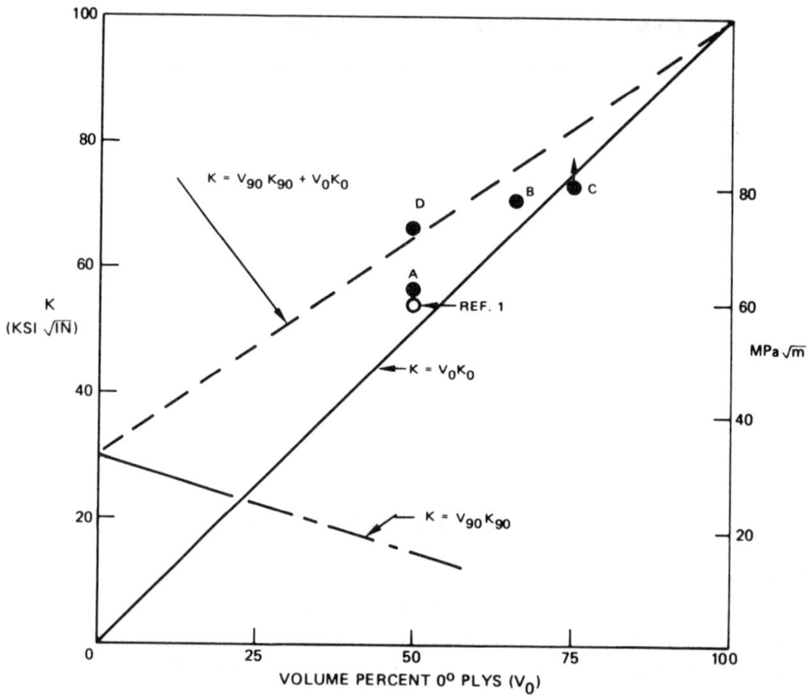

Figure 11. Maximum stress intensity of CT specimens for a/w = 0.6 as a function of percent of 0° plys.
Note: (Upper case letters refer to lay-up types)

lay-up which may relate to a stress concentration effect at the notch tip.

The dependence of composite fracture toughness on the ply lay-up sequence was shown to be significant. A comparison between composite Types A and D on the basis of stress intensity value at fracture and fracture morphology both point up this effect. The segregation of 0° and 90° plys caused a 20% increase in fracture toughness over the alternating ply lay-up sequence. This is quite different from a previously noted dependence of composite strength on lay-up sequence [6, 7], in that the unnotched composite tensile performance for the present study was totally unaffected by this change in sequence. The prior work, on resin matrix composites, was explained in terms of the generation of high interlaminar stresses due to the differences in Poisson's ratio between the various plys. During axial tensile loading the differential tendencies for transverse contraction generate transverse stresses which cause interlaminar failure. This occurs in both notched and unnotched specimens so that the smooth specimen tensile strength is affected as well as the notched tensile behavior. Clearly this explanation does not control the behavior of the current B/Al specimens. Instead the current data are best explained in terms of the restriction on matrix deformation and fracture in regions parallel to the 0° fiber direction. McGarry and Mandell [8] have demonstrated the significance of this effect in glass fiber reinforced plastic. In both the resin and metal matrix cases, the alternating ply lay-up sequence more effectively prevents crack tip blunting, while the segregated ply approach permits a greater amount of matrix deformation to take place in response to the transverse and shear stresses occurring at the tip of a crack. Because of the lack of any observable effect on composite unnotched tensile performance, the recognition of this type of behavior provides a significant opportunity for the achievement of enhanced structural reliability without penalty and adds another dimension to the process of composite design.

The difference between fracture toughness values obtained for CN and CT type specimens is not explained at this time. It probably relates to the observed dependence of K_{max} on (a/w) for the CT specimens wherein fracture toughness decreases with increasing (a/w). At larger values of (a/w) it may be that the two test techniques would agree; however, a/w of 0.6 is already quite large. In addition, the lay-up sequence evaluated contained alternating plys which should eliminate any thickness effect due to the high level of constraint at the crack tip. It is most likely that the discrepancy lies in the format used for calculation and that another approach should be used. This was not, however, within the scope of the current investigation.

REFERENCES

1. C. T. Sun and K. M. Prewo, "Fracture Toughness of B/Al," *J. Composite Materials*, Vol. 11 (1977), p. 164.
2. H. Tada, The Stress Analysis of Cracks Handbook, Del Research Corp., Hellertown, PA, 1973.
3. W. F. Brown and J. E. Srawley, "Plane Strain Crack Toughness Testing of High Strength Metallic Materials," in ASTM STP 410, 1966.
4. J. P. Waszczak, "Applicability of Linear Elastic Fracture Mechanics to 5.6B/6061," *J. Aircraft*, Vol. 13 (1976), p. 770.
5. J. Awerbuch and H. T. Hahn, "Hard Object Impact Damage of Metal Matrix Composites," *J. Composite Materials*, Vol. 10 (1976), p. 231.
6. N. J. Pagano and R. B. Pipes, "The Influence of Stacking Sequence on Laminate Strength," *J. Composite Materials*, Vol. 5 (1971), p. 50.
7. I. M. Daniel, et al., "Effects of Material and Stacking Sequence on Behavior of Composite Plates with Holes," *Exp. Mechanics* (1974), p. 1.
8. F. J. McGarry and J. F. Mandell, "Fracture Toughness Studies of Fiber Reinforced Plastic Laminates," Faraday Special Discussions of the Chemical Society, No. 2, 1972.

The Interlaminar Shear Strength of Unidirectional Boron-Aluminum Composites

P. W. M. PETERS

Deutsche Forschungs- und Versuchsanstalt für Luft- und Raumfahrt
Institut für Werkstoff-Forschung, Köln-Porz
West Germany

(Received August 24, 1977)

ABSTRACT

The interlaminar shear strength was estimated for four types of B-Al unidirectional composites. The shear strengths from the short beam shear test and symmetric block shear test sometimes show a considerable difference. It is shown that in the case of a matrix of high plasticity and (or) many defects in the shear-plane the short beam shear test gives a much too high shear strength. The symmetric block shear test supplies more realistic values both for the inter- and intralaminar shear strength.

INTRODUCTION

THE INTERLAMINAR SHEAR strength influences fracture and fatigue of composites and is an important factor in proving the quality of production. Several test techniques have been used for its determination [1, 2]. However, the state of pure shear is approached only in a few experiments, e.g. the torsion test. The complex state of stress, which occurs e.g. in the short beam shear test technique, can result in a mixed-mode shear/tensile fracture when fiber fracture takes place. Another problem is the different degree of shear stress concentrations which occur in the various test techniques. For these reasons it is difficult to compare the shear strength of the different test techniques. The short-beam shear test, introduced by ASTM as a standard method [3], is often used successfully, especially for glass and carbon fiber reinforced resins. Hanna et al. [4], however, showed that when the short beam shear specimen breaks in a complex failure-mode (tensile as well as shear failure), the resulting shear strength cannot even be a conservative estimate of the material's shear strength. Due to this imperfection of the short beam shear test method, another experiment, a symmetric block shear test, was used in the present investigation which is based on the particular shear behavior of unidirectional B-Al. The results of these symmetric block shear experiments are compared with the short beam shear strength. The shear strengths in the production plane (interlaminar) as well as perpendicular to this plane (intralaminar) are investigated.

MATERIALS

Boron-Aluminum of two different manufacturers was investigated. Material no. 1 consists of 145 μm B-SiC fibers in a matrix of 6061 (spray) and 4343 (foils), while material no. 2 contains 140 μm $B\text{-}B_4C$ fibers in the same matrix. The specimens were cut out of plates consisting of 20 unidirectional braze bonded tapes (process: 600°C, 10 at, 10^{-2} torr during 10 min).

A part of both materials was heat-treated for the T6 condition (cooled down from 525°C in water of 20°C, 2 h tempering at 165°C, 5–6 h tempering in a press at the same temperature). Due to the somewhat different plate thicknesses the fiber volume is different for the investigated materials:

1-T0: 64%
1-T6: 64%
2-T0: 51%
2-T6: 49%

The materials were brazebonded and heat-treated at MBB GmbH.

EXPERIMENTAL PROCEDURE

Short Beam Shear Test

The specimens were cut out of the different plates with a diamant wheel. They were of different length in order to make tests at different spans of 8, 12 and 15 mm. This resulted in the span to depth ratios for the interlaminar shear tests as shown in Table 1. The interlaminar shear tests at 12 and 15 mm span were performed with specimens of 19 mm length, so that they were considerably extending beyond the lower loading pins. To study the influence of these extending ends, specimens with length 13 mm were tested too on the 12 mm fixture. The specimens tested on the 8 mm fixture had a length of 9 mm. At midspan the deflection is measured with an extensometer and recorded against the load. Some additional intralaminar short beam shear tests were made for comparison with the symmetric block shear tests. In the latter case, the materials tested were 1-T0 and 2-T0 as shown in Table 1.

Symmetric Block Shear Test

In a block shear test a part of the specimen is sheared off under compression (Figure 1a). Due to the antisymmetric forces the specimens have to be supported. This is not necessary when the specimen is sheared off along two parallel planes (Figure 1b). When shear fracture occurs in both planes at the same time supporting

Table 1. Span to Depth Ratio of the Short Beam Shear Tests.

	Interlaminar Shear Tests			*Intralaminar Shear Tests*	
Span mm	*8*	*12*	*15*	*8*	*12*
Material					
1-T0	2,7	4,2	5,2	2,0	4,0
1-T6		4,2	5,2		
2-T0	2,4	3,6	4,6	2,0	4,0
2-T6		3,5	4,3		

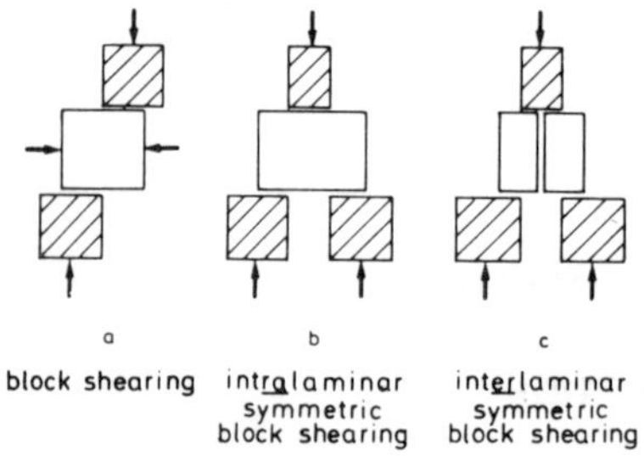

Figure 1. The block shear test technique.

of the specimen can be omitted. Thus the intralaminar shear strength is determined using 6 × 6 mm specimens, of which the middle part (width 3 mm) is sheared off (Figure 1b). The interlaminar shear strength was determined with the same size of specimen. In order to be able to shear off parallel to the surface two specimens are placed in the fixture surface to surface (Figure 1c). In all tests the displacement in the middle of the specimen was measured with an extensometer and recorded against the load. The critical parts of the fixture are made of cemented carbide.

RESULTS

Short Beam Shear Test

First the tests to determine the interlaminar shear strength are considered. Typical load deflection curves for the four materials are presented in Figure 2. After an elastic deflection in the beginning, the Al-matrix deforms plastically. Next irregularities, not drawn in the curves, occur caused by inhomogeneous matrix deformation (Portevin-Le Chatelier effect) or shear crack formation. Fiberfracture does not contribute to these irregularities. This was shown on some specimens, which were loaded until the first irregularity took place. After dissolving the matrix no broken fibers could be seen.

At higher loads fiberfracture do take place, resulting in bigger irregularities or a sudden load drop (Figure 1). Visible shear cracks, however, did not occur. The load-deflection curves clearly show that for the no. 1 materials the loading

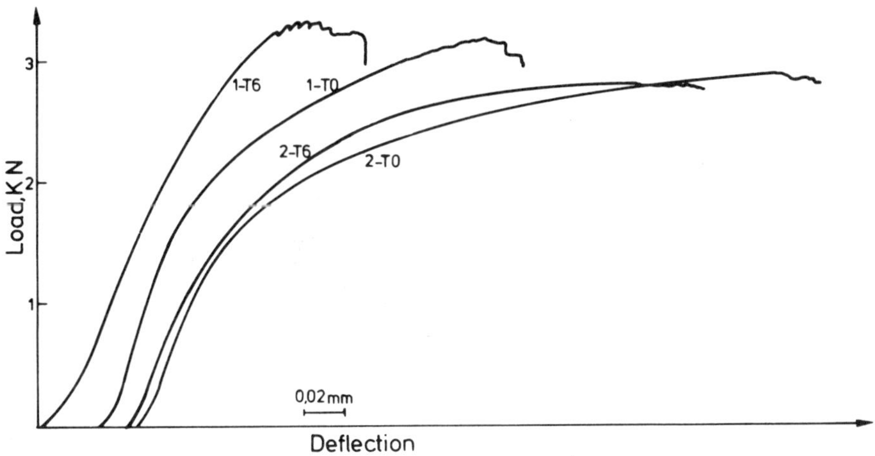

Figure 2. The load-deflection curves in short beam shear bending.

capability of the matrix is not yet reached, because irregularities due to fiber-fracture at the highly stressed midspan exists while the slope of the curve still is considerable positive. Fiberfracture of the no. 2 material occurs at a slope of zero or even at a negative slope, which shows, that the matrix is loaded to its limit. Fracture of the no. 2 material often took place at one of the lower supporting pins, when the specimen was extending beyond these pins. The other specimens showed fracture at midspan on the tensile side.

The intralaminar shear strength was investigated on some short beam shear specimens of material 1-T0 and 2-T0. The load-deflection curves for the 12 mm span specimens are of the type 1-T0 in Figure 2. Here again the specimens broke at last due to fiberfracture at midspan on the tensile side.

The inter- and intralaminar shear strength calculated by

$$\tau = \frac{3P}{4bd}, \text{ where:}$$

P = maximum load
b = specimen width
d = specimen thickness,

based on the elastic beam theory is presented in Figures 3 and 4.

Concerning the interlaminar shear strength, Figure 3 shows:

1. The apparent shear strength rises with decreasing span to depth ratio.
2. The no. 1 materials are more resistant to short beam shear bending than the no. 2 materials.

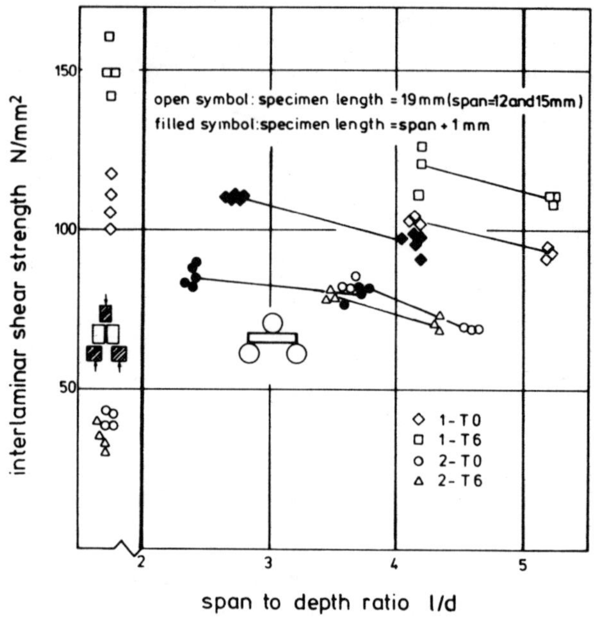

***Figure 3.** The interlaminar shear strength in short beam shear bending and symmetric block shearing.*

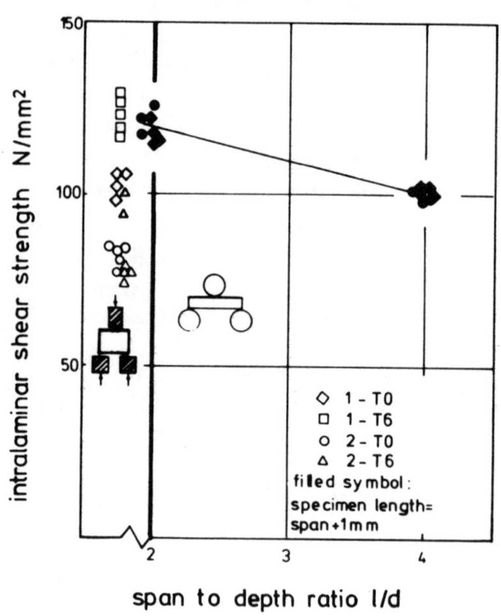

***Figure 4.** The intralaminar shear strength in short beam shear bending and symmetric block shearing.*

3. The heat treatment has a positive effect upon the interlaminar shear strength of the no. 1 material, while for the no. 2 material a reverse trend is visible.
4. There is a tendency to higher values when the specimen is longer than the span, especially for material no. 1.

The results of the intralaminar short beam shear tests on material 1-T0 and 2-T0 (Figure 4) show:

1. The shear strength again rises with decreasing span-depth ratio.
2. The intralaminar shear strength of 1-T0 and 2-T0 are about the same for both span to depth ratios.
3. Comparing the short beam intra- and interlaminar shear strengths it can be seen that for the no. 1 material the interlaminar shear strength has the highest values, while the reverse occurs for material 2-T0.

Symmetric Block Shear Test

The load-deflection curve of an intralaminar block shear test shows an elastic rise in the beginning, followed by plastic shear deformation (Figure 5). During this plastic deformation the load reaches a maximum and is continually decreasing until the middle part of the specimen is completely sheared off and pushed into the fixture gap, giving rise to a resistance load. This resistance was for most tests about 10–15% of the maximum load. When the resistance was higher the test results were considered invalid.

Sometimes the load fell abruptly to about the resistance load. This occurred, however, always after the load passed through the maximum. In the tests for the determination of the interlaminar shear strength this sudden load drop often occurred, caused by an abrupt shear crack extension in a bonding plane. An intralaminar shear crack, however, must find its way through the separate foils, which mostly results in slow crack extension and a gradually dropping load. Formation of the shear cracks takes place before the load has reached its maximum.

Seldom a shear crack was formed only on one side, which caused an antisymmetric stress distribution on the rest of the specimen in which case the test is useless. Normally, however, the differences in critical shear deformation between the two shear planes and (or) an unequal shear deformation due to inaccuracy of the specimen dimensions is compensated by the plastic deformation and the slow running shear crack.

The symmetric block shear strength is calculated with the maximum load and the area of the two shear surfaces. The results are presented in Figures 3 and 4.

They show:

1. Both inter- and intralaminar shear strengths of the no. 1 materials are higher than those of the materials no. 2.
2. The heat-treatment of material no. 1 has a positive effect on both inter- and intralaminar shear strengths, which is not the case for material no. 2.

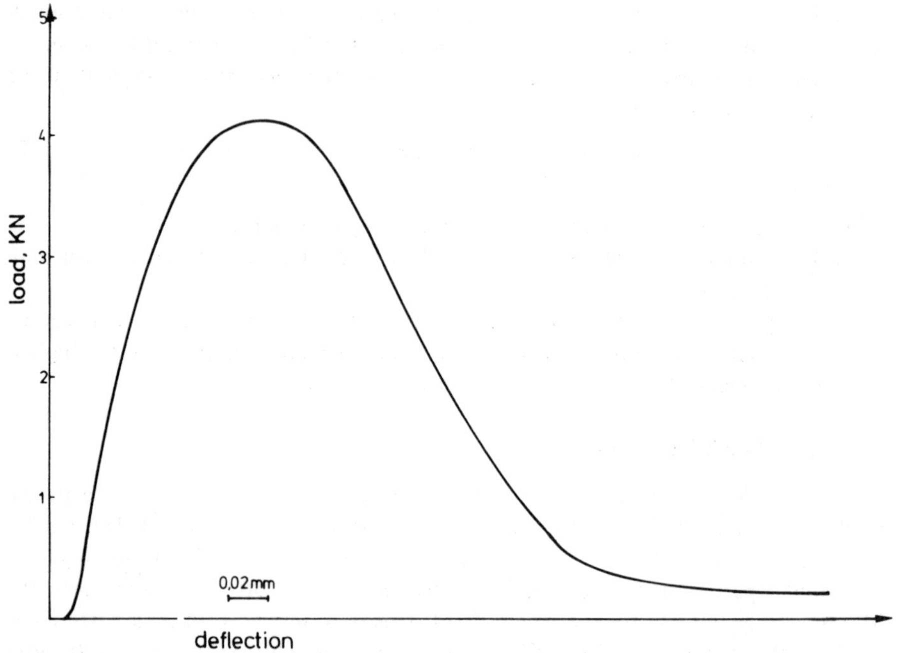

Figure 5. A typical load-deflection curve for a symmetric intralaminar block shear test.

3. For material no. 1 the highest shear strength is reached in interlaminar block shear testing, while for material no. 2 this occurs for intralaminar block shear testing.

METALLOGRAPHIC INVESTIGATIONS

A metallographic investigation of material 1-T6 shows a good bond between the different tapes with the 4343 foils clearly visible. The materials 2-T0 and 2-T6, however, proved to contain defects in the bonding plane. Throughout the investigated specimens small cracks were found between the separate tapes (Figure 6).

DISCUSSION

There is a considerable difference between the shear strengths of the used test techniques, mainly for the interlaminar shear strengths of the no. 2 materials. In this case the short beam shear test produces shear strength twice as high as the symmetric block shear technique does. The state of pure shear is better approached in the symmetric block shear test, where shear fracture always occurred in contrast to the short beam shear test, where the specimen finally broke due to fiber rupture.

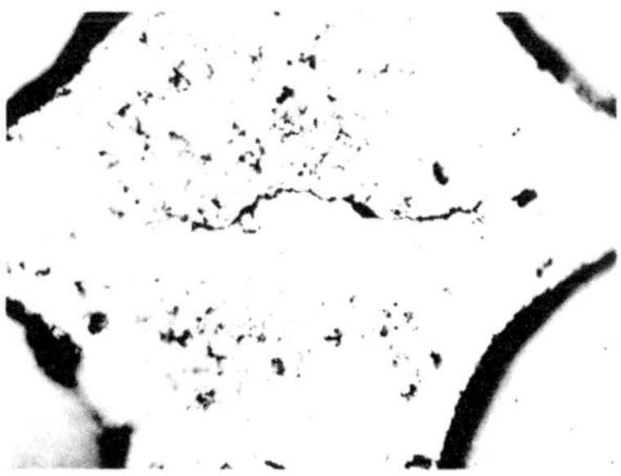

Figure 6. Cross-section of material 2-T6 showing a shear crack in the bonding plane.

Therefore it is the short beam shear strength that has to be doubted. Hanna et al. [4] came to the same conclusion before. They concluded so, after comparing shear rod with short beam shear tests of graphite epoxy composites, when in short beam shear testing fracture with tensile components occurred.

One reason for a too high value of the short beam shear test can be the plastic deformation of the matrix (see Figure 2), which in effect makes the shear stress calculation based upon the elastic beam theory invalid. The flattening of the load deflection curve is mainly caused by a plastic shear deformation of the matrix and only to a minor extent by the plastic deformation of the matrix in tension and compression. This influences the shear stress distribution in the specimen, which based on the linear elastic theory is a parabola with maximum shear stress in the middle part of the specimen. Due to the plastic shear deformation this parabola flattens, while near the outer surfaces the shear stress increases. Thus, the linear elastic beam theory produces too high shear values. When the matrix shows ideal elastic-plastic behavior and fiber rupture does not occur the shear stress can approach the theoretical value of $\tau = P/2bd$, which is 33,3% lower than the value of the elastic beam theory.

Berg et al. [5] showed that locally the shear stress can rise as high as twice the maximum stress of the elastic beam theory. Using a finite element analysis they calculated the shear stresses for a graphite-epoxy composite with a matrix with slightly strain-hardening properties. The highest shear stresses occur in the vicinity of the upper loading nose. Apart from occurring locally near the pins, this shear increasing effect also takes place before shear yielding, starting from the specimen mid-shearplane, causes the shear redistribution described before. These two effects are conflicting and it depends on the elastic-plastic behavior of the matrix and the specimen dimensions, how they influence the shear stress.

The about twice as high interlaminar short beam shear strength of materials no. 2 (compared with the symmetric block shear strength) cannot be explained with these phenomena. The main cause for the exaggerated high shear strength must lie in the fact that there is a compressive stress at the shear plane. Berg et al. [6] reported a rising shear strength when under biaxial compression the shear cracks are closed and the shearing ductility of the matrix is improved. Also under uni axial compression on the slip-plane a higher shear strength can be found [7]. This phenomenon contributes to the high interlaminar short beam shear strength of material 2-T0 and 2-T6, which contain many defects.

The rising shear strength of these materials at decreasing span to depth ratio must be explained with an increasing influence of these compressive stresses at the smaller span as the fracture mode and load-deflection curve does not essentially alter.

In intralaminar short beam shear bending the shear strength of material 1-T0 (at $l/d = 2$) and 2-T0 (at $l/d = 2$ and 4) is higher than the symmetric block shear strength, although the difference is less aggravating. As the short beam shear specimen breaks due to fiberfracture at a positive slope of the load-deflection curve the difference can grow bigger, when the shearing capability is completely exhausted.

The high plasticity of the matrix causes that the short beam shear specimen breaks due to fiber rupture instead of shear crack formation. This plasticity, however, favors the symmetric block shear test. Differences in shear behavior in the two shearing planes and inaccuracies in the specimen dimensions are compensated by this plasticity and the slowly growing shear cracks. The state of stress is not purely shear because flexure stresses exist in case of the intralaminar shear test. In some experiments with two smaller specimens side by side, as used for the interlaminar shear strength determination, a good agreement with the 6 × 6 mm specimens was found. This justifies the conclusion that the flexure stresses are small. A disturbing factor is the resistance exerted on that part of the specimen that is pushed into the fixture. In most cases, however, this resistance, after total shear-fracture occurred, was about 10–15% of the maximum load. At maximum load the deflection is about half as big; so the frictional force can be estimated to be smaller than 10% of the maximum load.

The metallographic investigations make clear that the brazebonding process for the no. 2 materials is insufficient, because small cracks occur in the bonding plane (Figure 6). This explains the low interlaminar shear strength. The intralaminar shear strength is, however, less effected by this bad bonding. As can be expected for such a badly bonded material the heat treatment does not improve the interlaminar shear strength. The intralaminar shear strength could be effected positively, which can not be confirmed clearly by the test results. The heat treatment of the no. 1 material improves both the inter- and intralaminar shear strength. This fact indicates a low number of defects in the brazing zone and justifies the conclusion that the bonding is good.

CONCLUSIONS

1. The short beam shear test is not useful when the composite contains a matrix of high plasticity which causes the fracture mode to be controlled by fiber fracture. This plasticity of the matrix results in a shear stress redistribution in the specimen when the shear yield stress is exceeded. This can cause an overestimated shear strength value (max. 33,3%). Besides, when the shear strength is limited by defects, local compressive stresses near the loading pins at the shear plane diminish the influence of these defects. This phenomenon is most important for a small span.

2. When the matrix has a high plasticity the described symmetric block shear test is useful. Conditions, however, are a continuous load-deflection curve (as the result of a plastic matrix deformation and the formation of slow running shear cracks in both planes) and a not too high frictional force, which is necessary to push the sheared-off part into the fixture.

REFERENCES

1. G. Menges and R. Kleinholz, "Vergleich Verschiedener Verfahren zum Bestimmen der Interlaminaren Scherfestigkeit," Kunststoffe, Vol. 59 (1969), p. 959.
2. R. F. Zabora and J. E. Bell, "A Test Technique to Study Interlaminar Shear Phenomena of Laminated Composites," Technical Report AFFDL-TR-71-67 (1971), (AD-729 872).
3. Apparent Horizontal Shear Strength of Reinforced Plastics by Short Beam Method, ASTM D 2344-76.
4. G. L. Hanna and S. Steingiser, "Defining the Adhesion Characteristics in Advanced Composites," ASTM STP 460 (1969), p. 182.
5. C. A. Berg, J. Tirosh and M. Israeli, "Analysis of Short-Beam Bending of Fiber Reinforced Composites," ASTM STP 497 (1972), p. 206.
6. C. A. Berg, R. Melton, I. Kalnin and T. Dunn, "Fracture in the Grip Section of Fiber Reinforced, Reduced Cross Section Tension Specimens," *J. of Mat. JMLSA*, Vol. 6 (1971), p. 683.
7. Berechnung und Versagensvorhersage von Faserverbundstrukturen. *ZTL-Zukunfts-Technik-Luft*, Fag 2, Jahresabschluβ-bericht 1975.

K-Calibration of Unidirectional Metal Matrix Composites

JONATHAN AWERBUCH

Department of Mechanical Engineering and Mechanics
College of Engineering, Drexel University
Philadelphia, PA 19104

AND

H. THOMAS HAHN

Nonmetallic Materials Division
Air Force Materials Laboratory
Wright-Patterson AFB, Ohio 45433

(Received May 17, 1978)

ABSTRACT

Load-displacement curves were obtained for center notch unidirectional boron/aluminum and borsic/titanium composites. Far field displacements, obtained by means of two standard clip gages, 1 and 0.25 inch long, respectively, were used to establish the variation of *global* compliance with crack length. The *local* compliance curve was obtained from the crack opening displacement (COD) measured across the crack surfaces by means of the laser interferometric technique. The Interferometric Displacement Gage (IDG), which is more sensitive to the appearance of damage at the crack tip, yielded nonlinear load displacement curves and seemed to better characterize the fracture behavior of composites. The experimental compliance curves and the *K*-calibration results were found to be in good agreement with analytical predictions. The *K*-calibration curves of the composites studied were found to be almost equal to that of the isotropic materials; both compliances were predictable by using the isotropic *K*-calibration.

SCOPE

An important parameter in the analysis of linear elastic fracture mechanics is the finite width correction factor, i.e. the *K*-calibration factor. It has been shown analytically [1] that for anisotropic materials the *K*-calibration factor depends on the elastic properties. This fact introduces an additional complexity in the study of fracture behavior of composites, which are mostly anisotropic, because each composite material requires its own K-calibration factor. However, at least for the center crack geometry, it has been shown [2] that the error in using the isotropic values of the K-calibration factor for anisotropic materials decreases with increasing specimen length-to-specimen width ratio although it increases with increasing crack length-to-specimen width ratio. However, when the specimen length is more than triple the width, the error reduces to less than 5% for crack lengths up to half the specimen width [2]. Thus, the isotropic K-calibration factor can be used with reasonable accuracy when the proper geometry is chosen.

In the present paper we present experimental load-displacement curves from which the change in compliance with crack length can be measured. Two different test procedures, interferometric displacement gage and standard clip gages, are employed to measure the crack opening displacement and the far field displacement, respectively. The *local* and *global* compliances are obtained from these measurements and subsequently are compared with analytical predictions based on the isotropic K-calibration factor.

The materials chosen are unidirectional boron/aluminum (B/Al) and two types of borsic/titanium (BSiC/Ti) composites. The present work was motivated by a recent study on hard object impact damage of metal matrix composites [3]. In this study it has been established that the damage inflicted by the impact of a hard object can be considered to be equivalent to a through-the-thickness crack for the purpose of residual strength prediction. Thus, a study on the effect of lateral damage on the fracture behavior of metal matrix composites has been pursued with specimens in which center cracks were artifically introduced.

CONCLUSIONS AND SIGNIFICANCE

The laser interferometric technique has been successfully employed to measure the crack-opening-displacement (COD) of center-notched unidirectional metal matrix composites. It was found that this technique is highly sensitive in detecting crack tip damage, very accurate in the submicron scale, and simple to utilize. For comparison, load-COD curves were also obtained by using standard clip gages.

The load-COD curves for center-notched unidirectional B/Al composites obtained with the IDG technique are highly nonlinear. The load-COD curves for unidirectional BSiC/Ti composites exhibited a smaller degree of nonlinearity. Due to this nonlinearity, the local compliance values were measured from the unloading slopes of the load-COD curves. A comparison of the experimentally obtained local compliance curve with the prediction based on elastic stress analysis shows an excellent agreement.

The initial slopes of the load-COD curves obtained with the standard clip gages are linear, as expected, for all materials tested. The prediction of global compliance based on the energy release rate approach agrees very well with the experimental results.

The local compliance curve is more sensitive to the crack tip damage than the global compliance curve, indicating that the IDG technique is more accurate in studying the fracture behavior.

In the predictions of the local and the global compliance curve, the analytical isotropic *K*-calibration factor was used. Results of the *K*-calibration factor obtained from the local complance curve show that the isotropic values can be used for crack length-to-specimen width ratios greater than 0.1. Since the degree of anisotropy is less with multidirectional composite laminates, the same conclusion can be extended to a wider class of composites.

EXPERIMENTAL PROCEDURE

Interferometric Displacement Gage (IDG)

The crack opening displacement (COD) was measured by means of the laser interferometric technique which enables measuring displacement over a very short gage length. When a coherent monochromatic light from a laser source impinges upon two closely adjacent, reflecting indentations on the sample, it is reflected (diffracted) back at an angle α_o with respect to the incident beam, Figure 1. The diffracted beams overlap, generating interference fringe patterns on each side of the incident laser beam. As the distance (d) between the indentations changes due to the applied load, the two fringe patterns move. Observed from a fixed position at an angle α_o, this movement can be related to the change in distance between the indentations, δd.

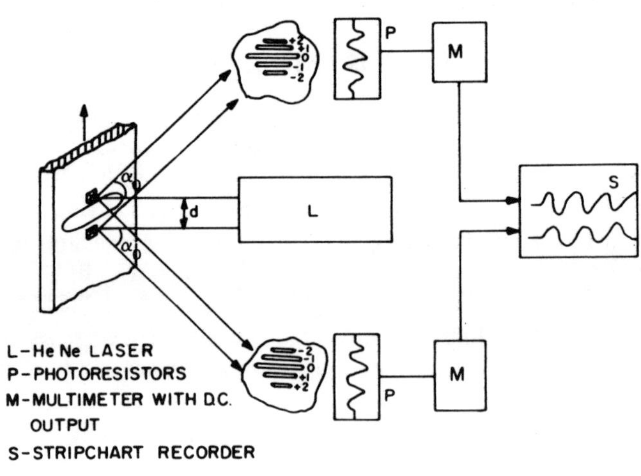

Figure 1. Schematic view of the laser interferometric displacement technique. Nominal value of d is 0.5 mm and α_o is typically 42°

The equation locating the bright interference fringes is [4]

$$d \sin\alpha_o = m\lambda \tag{1}$$

where m is the fringe order ($m = 0, \pm 1, \pm 2, \pm 3 \ldots$), λ is the wave length of the laser beam, d is the spacing between indentations and α_o is the angle between the incident and the reflected beam.

The change in distance between indentations, i.e. the displacement δd, is related to the change in fringe order, δm, at the fixed observation point by

$$\delta d = \frac{\delta m \lambda}{\sin\alpha_o} \tag{2}$$

Note that δm is the number of fringes (or fraction of fringes) passing the fixed observation position.

Using a laser beam wavelength λ of 0.6328 μm (He-Ne laser) and $\alpha_o \approx 42°$ the calibration constant $\lambda/\sin\alpha_o$ in Equation (2) is approximately 0.95 μm. Thus, a fringe motion of $\delta m = 1$ corresponds to a displacement of about one micron. Since the shift of one half fringe movement is easily detectable, the resolution of the measurement easily reduces to a submicron scale. The displacement range in this technique is limited and depends on the spacing between indentations, dimensions of the indentations, and the power of the recording system. However, a displacement range of up to 300 μm was easily obtained in our experiments.

In order to obtain clear fringe patterns, the specimens should have a fairly flat and smooth surface. As a final preparation, the specimens were polished with a 3 μm diamond paste. A shallow, reflective indentation of a pyramidal shape was applied to the specimen surface on each side of the crack at the center with a Lietz microhardness tester. The indentation dimensions were approximately 50 X 50 μm with edges normal and parallel to the crack surfaces as shown in Figure 2, and the original distance between the indentations was about 500 μm.

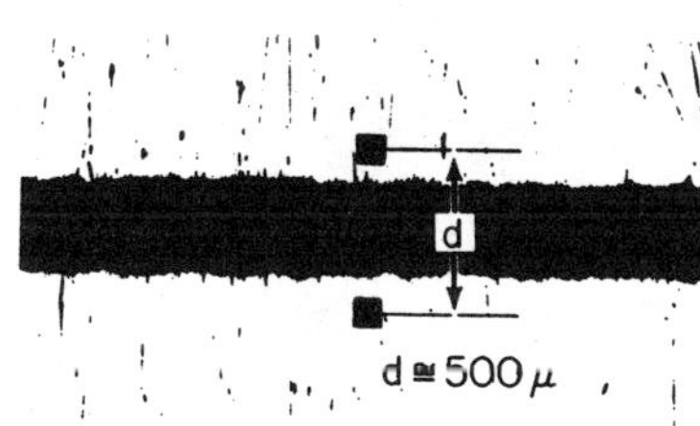

Figure 2. Photomicrograph of two indentations located at the center of notch.

The instrumentation system for measuring the crack opening displacement is simple and economical, and consists of a Spectra Physics Model 134, 3mW He-Ne laser, two photoresistor detectors, two signal amplifiers with d-c voltage output, and a dual channel stripchart recorder, as shown schematically in Figure 1. The sensing elements were two CdS bulk effect photoresistors. The active face of each of them was masked, leaving a 0.1 mm slit which was narrower than the fringe spacing so that the photoresistors responded to well defined maximum and minimum values of intensity. The output signals of the photoresistors were amplified and recorded on a stripchart recorder. This recording technique is essentially similar to the one described in Reference [5]. A comprehensive description of the laser interferometric technique is given in Reference [6].

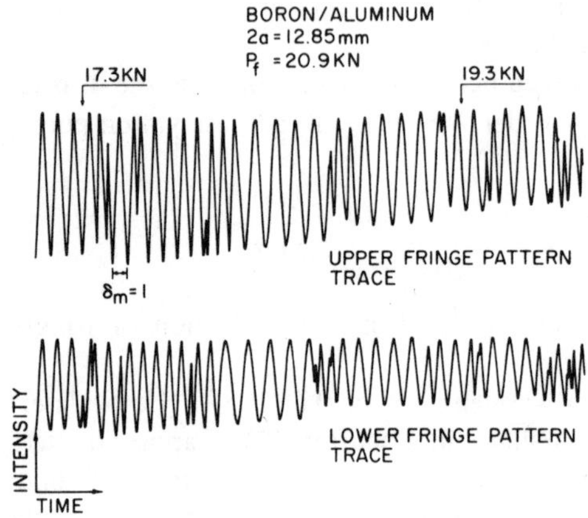

Figure 3. *Typical strip chart recorder trace of fringe motion prior to failure.*

Figure 4. *Photograph of compliance gage attached to specimen (gage length = 1.0 inch).*

Figure 3 shows a typical intensity plot obtained on the stripchart recorder as the fringe patterns pass by the two stationary photoresistors. This particular plot is for a boron/aluminum specimen having a 12.85 mm (0.5 in.) crack. The data reduction of these signals is fairly easy; correlating the time at which various maxima and minima occur ($\delta m = 0, 1/2, 1, 1\text{-}1/2 \ldots$) with the load-time plot obtained from the Instron chart recorder yields the required load-COD curve. Fringe motion occurs with any rigid body motion of the sample. However, rigid body motion can be eliminated by averaging the displacement obtained from the upper and lower photoresistors, Figure 1 [6].

Clip Gages

As the Interferometric Displacement Gage (IDG) technique has not been used before for composite materials, it was also of interest to compare the load-COD curves with the load-displacement curves obtained using the standard clip gages.

Two types of gages were used in this study. The first with a gage length of 1 in., hereafter called the compliance gage, was attached to the specimen by friction with a pair of knives on each side of the crack at the center of the specimen, Figure 4. The second, called the clip gage, was a double centilever clip gage similar to the one used in fracture testing of metals [7]. This clip gage was attached to steel tabs (0.25 X 0.25 in.) which were bonded to the specimen with epoxy cement 0.25 in. apart, Figure 5.

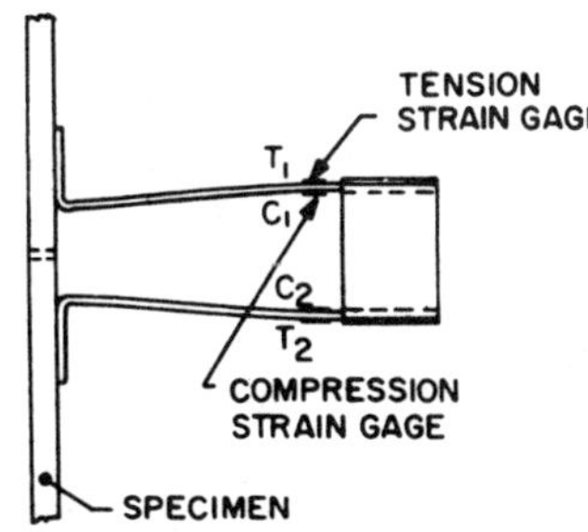

***Figure 5.** Schematic view of clip gage (gage length = 0.25 inch).*

Materials

The materials tested in this program are given in Table 1. All specimens were cut from 5.5 X 5.5 in. plates, parallel to the fiber direction, with a diamond saw or by E.D.M. (Electric Discharge Machining) method, and were ground on the edges to ensure minimum damage. The specimen dimensions

Table 1. Materials Tested[(a)].

Composite	Matrix Material	No. of Plies	Thickness	Fabrication Procedure	Manuf.
B/Al	Al-6061-F	8	0.052	Air Bonding	United Tech.
BSiC/Ti-A	Ti-6-4	6	0.052		
BSiC/Ti-B	Ti-6-4[(b)]	6	0.068	Diffusion Bonding	T.R.W.

(a) Boron fibers are 0.056 in. in diameter, 48-52 V/o

(b) Specimens have an extra 0.012 in. thick layer of Ti-3Al-2 1/2 V on each surface.

were 5.5 in. long by 1.0 in. wide. Center notches of about 0.05, 0.1, 0.2, 0.3, 0.4 and 0.5 in. long and 0.01 in. wide were introduced by E.D.M. The total number of center-crack tension specimens tested in this program includes 16 specimens of B/Al, 14 specimens of BSiC/Ti-A, and 15 specimens of BSiC/Ti-B. From each plate one specimen without a notch was tested statically to provide base line strength data. the Average fracture strength and longitudinal stiffness were 200.4 MPa 200, 4 Mpa (290.7 ksi) and 241.3 Gpa (35 X 10^3 ksi), respectively, for the B/Al specimens, 832.2 MPa (120.7 ksi) and 237.9 GPa (34.5 X 10^3 ksi), respectively, for the BSiC/Ti-A specimens, and 812.2 MPa (117.8 ksi) and 204.1 GPa (29.6 X 10^3 ksi), respectively, for the BSiC/Ti-B specimens. Aluminum doublers of 1 in. square were sheared from 0.05 in. thick 2024-T6 sheets and were glued with epoxy onto the specimen ends. Both the aluminum doublers and the ends of the composite specimens were etched with a diachromatic acid solution for 10 minutes at 140°F. The adhesive material consisted of aluminum matrix epoxy containing 90% aluminum by weight.

RESULTS AND ANALYSIS

The laser interferometric and the clip gage techniques described in the preceding section were used to obtain load-COD curves from which the experimental compliance and *K*-calibrations for the center-notched unidirectional B/Al and BSiC/Ti composites could be calculated.

A typical load-COD curve obtained from the interferometric displacement gage (IDG) is shown in Figure 6 for a B/Al specimen having an initial crack length of 12.85 mm (0.5 in.). The load-COD curve is highly nonlinear followed by rapid increases in COD when the load approaches the fracture strength. The rapid changes in COD are due to fiber failures and also to longitudinal plastic deformation and splitting in the matrix [8]. The damages were also accompanied by audible levels of acoustic emission. A detailed examination of the load-COD curves reveals that the load-COD relationship remains highly nonlinear also between loads where there is a rapid increase in COD. A detailed discussion of the damage growth at the crack tip, fracture strength, modes of failure, and experimental resistance curves for B/Al is given in Reference [8].

The high nonlinearity of load-COD curves makes it difficult, and practically impossible, to accurately measure the elastic compliances, and consequently, to determine the stress intensity calibration curve. Therefore, in using the IDG technique, each specimen has been loaded to a predetermined load and then unloaded. This procedure was repeated several times incrementally up to failure as shown in Figure 6. An examination of the load-COD curve in Figure 6 reveals that the unloading curve is initially linear just as is the subsequent loading curve. It was this

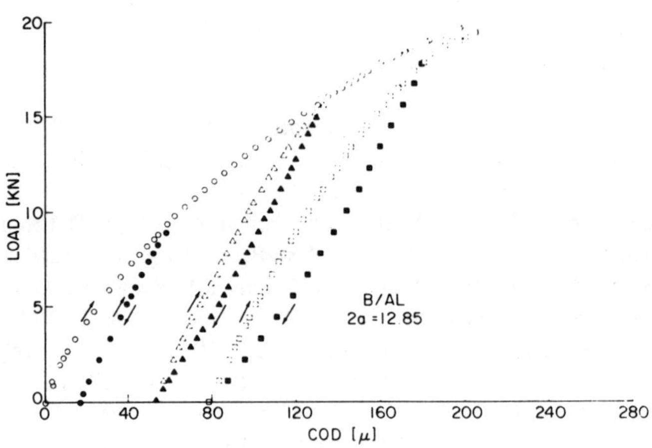

Figure 6. Load-COD (loading-unloading) curves obtained with interferometric displacement gage for boron/aluminum (crack length = 12.85 mm).

linear portion of the load-COD curve that was used to obtain the compliance as a function of the initial crack length. It should be noted that the initial slopes of the repeated loading and unloading curves were essentially the same, Table 2. However, as the load approaches its ultimate level, a notable increase in compliance is observed, Table 2, due to excessive damage at the crack tip. Figure 6 also shows that upon unloading a relatively large permanent displacement is obtained. After each unloading the specimens were removed from the Instron machine and the permanent opening of the crack was directly measured under a microscope. A comparison of the permanent displacements measured from the load-COD curves and from the direct measurements shows an excellent agreement, demonstrating the accuracy of the IDG technique.

Table 2. Experimental Values of Local Compliance Measured from Loading/Unloading-COD Curves

B/AL. Crack length = 7.47 mm (0.294 in.)

load level [N (lb)]	360 (1600)		450 (2000)		1124 (5000)		1349 (6000)		1461 (6500)-Failure	
	Loading	Unloading	Loading	Unloading	Loading	Unloading	Loading	Unloading	Loading	Unloading
Compliance [10^{-3} μm/N]	2.56	2.49	2.53	2.45	2.42	2.55	2.55	2.83	2.75	

BSiC/Ti-A: Crack Length = 7.49 mm (0.295 in.)

load level [N (lb)]	450 (2000)		730 (3250)		787 (3500)		843 (3750) Failure	
	Loading	Unloading	Loading	Unloading	Loading	Unloading	Loading	Unloading
Compliance [10^{-3} μm/N]	2.56	2.49	2.38	2.53	2.49	2.56	2.54	

Both the nonlinearity of the load-COD curve obtained with the IDG and the large permanent COD indicate the occurrence of crack-tip damage even at relatively small loads. However, this damage is subcritical and does not affect the compliance measurements, Table 2. Figure 6 also shows the "combined" load-COD curve, constructed by combining the individual loading-unloading COD curves. It is seen that the overall load-displacement behavior is very similar to the one for metals.

Similar load-COD curves were also obtained for BSiC/Ti, Figure 7, for which the nonlinearity is not so pronounced as for B/Al. Generally, the fracture behavior and the modes of failure of BSiC/Ti composites are quite different from those of B/Al; the difference manifests itself also in the "combined" load-COD curves for the two materials, as shown in Figure 8. A detailed discussion of this aspect is given in Reference [9].

A comparison between the load-COD curves from the IDG and the load-displacement curves both from the clip gage and from the compliance gage is shown in Figure 9. As expected, the load-displacement curves obtained with the clip and the compliance gage are initially linear, followed by jumps in displacement at higher load levels. However, a detailed examination of the load-displacement curves also indicates that the load-displacement relationship remains linear even between loads

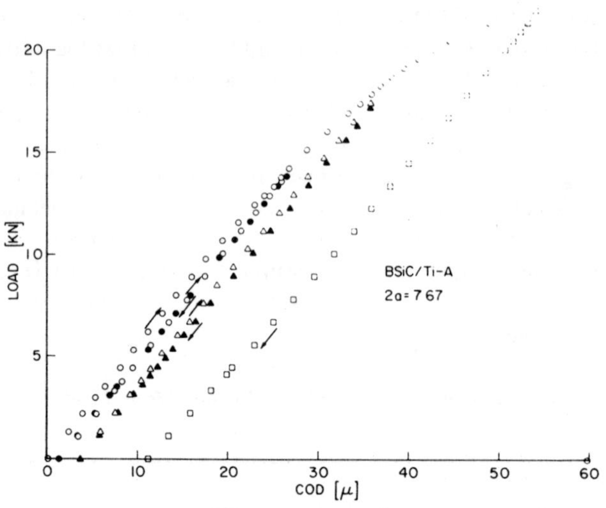

***Figure** 7. Load-COD (loading-unloading) curves obtained with interferometric displacement gage for borsic/titanium-A (crack length = 7.67 mm).*

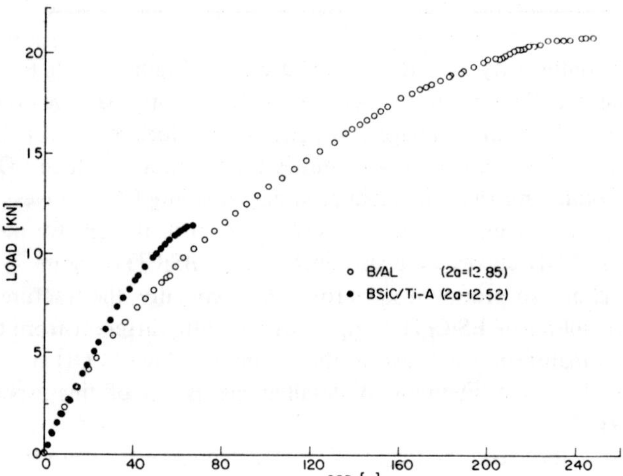

***Figure** 8. Comparison of load-COD curves obtained with interferometric displacement gage for boron/aluminum and borsic/titanium-A (both of similar crack length).*

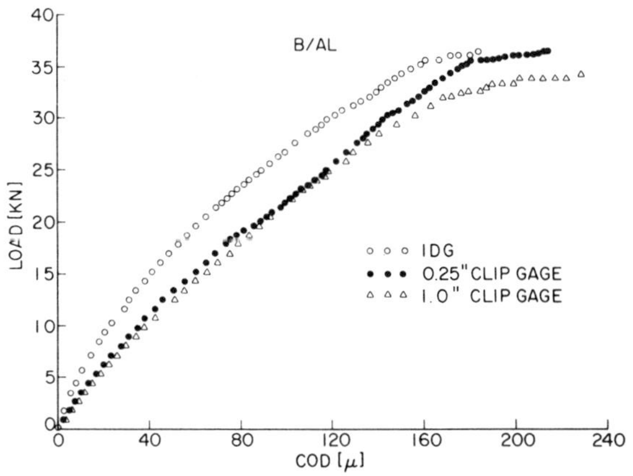

Figure 9. Load-COD curves for boron/aluminum (crack length = 12.85 mm) obtained with interferometric displacement gage, compliance gage, and clip gage.

at which there is a jump in displacement. Similar behavior has been observed in other works, e.g. [10], [11] and [12], for various resin and metal matrix composites, using various types of measuring techniques. In fact, only with the IDG technique, a continuously nonlinear load-COD curve is observed. For several specimens the strain at the crack-tip was recorded using a 1/16 in. strain gage. Again, the load-strain records were initially linear, as was observed in Reference [12]. Basically, the IDG demonstrates itself to be highly sensitive in detecting crack-tip damage.

From both test procedures (IDG and clip gages), the compliance calibration as a function of the initial crack length was obtained. Since two different displacements were measured, appropriate compliances were defined accordingly: the *local* compliance, C_ℓ, based on the COD, and the *global* compliance C_g based on the far field displacement. The local compliance was obtained from the loading-unloading COD curves as listed in Table 2. The compliance values given in Table 3 are the average values of 3 to 6 loading-unloading curves. The global compliances were measured directly off the load-displacement curves obtained from the compliance gage, and are also listed in Table 3.

The application of the energy release rate approach can predict the global compliance. Consider a crack of length 2a in a unidirectional composite plate, Figure 10. The fibers are in the y-direction and the crack is perpendicular to the fibers. The uniform far field stress is denoted by σ.

The energy release rate due to crack extension, which can be found in [13], is given by:

$$G = \alpha \frac{K^2}{E_L} \tag{3}$$

Table 3. Experimental Values of Local (C_ℓ) and Global (C_g) Compliance [10^{-3} μ/N]

Boron/Aluminum

Crack Length mm (in.)	C_ℓ	Crack Length mm (in.)	C_g
1.27 (0.05)	0.693	1.35 (0.053)	3.43
2.64 (0.104)	0.953	2.64 (0.104)	2.92
5.03 (0.198)	1.738	5.23 (0.206)	3.39
7.47 (0.294)	2.494	7.59 (0.299)	3.77
7.72 (0.304)	2.57	-	-
10.18 (0.401)	3.58	10.18 (0.401)	4.31
12.85 (0.506)	4.53	12.85 (0.506)	5.87

Borsic/Titanium-A

Crack Length mm (in.)	C_ℓ	Crack Length mm (in.)	C_g
1.1 (0.043)	0.564	1.37 (0.054)	3.04
2.43 (0.095)	1.00	2.69 (0.106)	2.78
5.23 (0.206)	1.82	5.26 (0.207)	3.77
7.49 (0.295)	2.52	7.54 (0.297)	3.81
10.08 (0.397)	3.45	-	-
10.27 (0.404)	3.59	10.13 (0.399)	4.12
12.52 (0.493)	4.68	12.80 (0.504)	5.65

Borsic/Titanium-B

Crack Length mm (in.)	C_ℓ	Crack Length mm (in.)	C_g
1.17 (0.046)	0.494	1.32 (0.052)	2.44
2.62 (0.103)	0.88	2.62 (0.103)	2.86
5.23 (0.206)	1.53	5.23 (0.206)	2.86
7.24 (0.285)	2.11	7.49 (0.295)	3.61
7.67 (0.302)	1.82	-	-
10.03 (0.395)	3.0	10.24 (0.403)	3.68
11.64 (0.458)	4.13	12.75 (0.502)	4.53

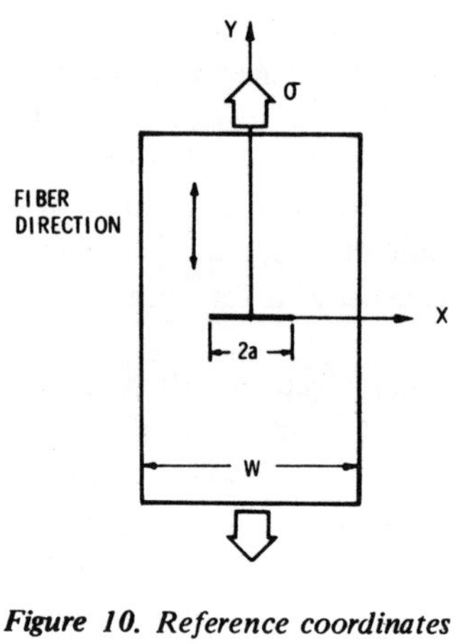

***Figure 10.** Reference coordinates for the center notched specimen subjected to uniaxial loading.*

where α is the orthotropy correction factor

$$\alpha = \frac{1}{\sqrt{2}} \left[\left(\frac{E_L}{E_T} \right)^{1/2} - \nu_{LT} + \frac{E_L}{2G_{LT}} \right]^{1/2} \tag{4}$$

and K is the stress intensity factor

$$K = Y \sigma \sqrt{\pi a} \tag{5}$$

In the foregoing equations, E_L and E_T are the longitudinal and the transverse modulus, respectively, ν_{LT} the major Poisson's ratio, and G_{LT} the longitudinal shear modulus. Y is the K-calibration factor.

The same energy release rate can be expressed in terms of the change in the *global* compliance C_g:

$$G = \frac{1}{2} \frac{P^2}{BW} \frac{dC_g}{d(2a/W)} \tag{6}$$

where B and W are the thickness and width, respectively, of the plate. Equating Equation (6) with Equation (3) and integrating the resulting equation, we obtain:

$$C_g = \frac{\alpha}{E_L B} \int Y^2 (a) \left(\frac{2a}{W} \right) d \left(\frac{2a}{W} \right) + M \tag{7}$$

where M is the integration constant, which is simply the compliance in the absence of a crack.

If the isotropic values are applicable for Y [7], i.e.

$$Y = 1 + 0.1282(2a/W) - 0.2881(2a/W)^2 + 1.5254(2a/W)^3 \tag{8}$$

then the final form for C_g becomes:

$$C_g = \frac{\alpha}{E_L B} [1.566(2a/W)^2 + 0.268(2a/W)^3 - 0.438(2a/W)^4 + 1.865(2a/W)^5 + 0.247(2a/W)^6 - 0.393(2a/W)^7 + 0.911(2a/W)^8] + \frac{L}{BWE_L} \tag{9}$$

where L is the distance between the far field reference points. Here L is taken as the length of the clip or compliance gage.

In Equation (9) the orthotropy correction factor α can be determined from Equation (4) once the engineering constants are given. The values of α for the three materials tested in this program were determined earlier [3] : $\alpha = 1.303$ for B/Al: $\alpha = 1.247$ for BSiC/Ti-A; and $\alpha = 1.247$ for BSiC/Ti-B. The longitudinal stiffness E_L was 241.3 GPa (35 $\times$ 10^3 ksi) for B/Al, 237.9 GPa (34.5 $\times$ 10^3 ksi) for BSiC/Ti-A, and 204.1 GPa (29.6 $\times$ 10^3 ksi) for BSiC/Ti-B. Note that in using Equation (9), the distance L is 1 in. for the compliance gage and 0.25 in. for the clip gage.

The predicted *global* compliance curve based on the isotropic Y has been compared with the experimental results obtained from the compliance gage measurement and an excellent agreement is established as shown in Figure 11 for all three materials. Surprisingly, the clip gage data do agree with the predictions as well even though a different integration constant M should be used due to the different gage length. However, it should be remembered that the gage length for the 0.25 in. clip gage is not clearly defined due to the effect of the tabs on the clip gage reading.

The interferometric displacement gage (IDG) enables measuring the crack opening displacement (COD) very close to the crack surfaces (gage length of 0.02 in.). This COD can be predicted from the elastic stress analysis, and hence the *local* compliance curve can be calculated. By using the local compliance C_ℓ, the applicability of the isotropic Y can also be checked. To this end we first note that the displacement v in the y direction of the top crack surface is given by [14].

$$v = \frac{2\alpha}{E_L} Y \sigma \sqrt{a^2 - x^2} \tag{10}$$

Since the COD is the relative displacement at the center of the crack, which is equal to 2v, the local compliance becomes:

$$C_\ell = \frac{COD}{\sigma BW} = \frac{2\alpha}{E_L B} \frac{2a}{W} Y \tag{11}$$

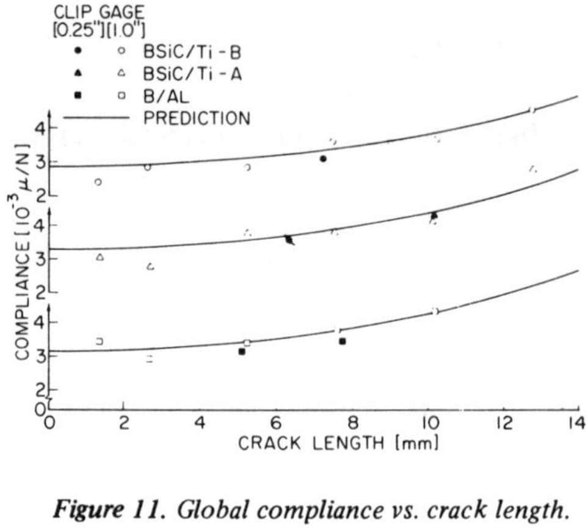

Figure 11. Global compliance vs. crack length.

The foregoing equation is compared with the experimental data in Figure 12. Again, an excellent agreement is observed.

A comparison between the global and the local compliance, Figures 11 and 12, shows that the latter is much more sensitive to the change in crack length. The higher sensitivity of the local compliance curve enables one to determine more accurately the size and the effect of an unknown pre-existing damage by measuring the compliance. The compliance curve may also serve as a tool for calculating the crack growth resistance

$$K_R = Y\sigma\sqrt{\pi a_e} \tag{12}$$

where a_e is the effective crack half length which yields the measured compliance on the compliance calibration curve. In other words, the original crack half length is replaced by a_e through the compliance matching as the crack tip damage develops.

A comparison between the experimental results and Equation (8) for Y is shown in Figure 13. The experimental values of Y can be obtained by rewriting Equation (11) as follows:

$$Y = \frac{E_L B}{2\alpha} \frac{C_\ell}{2a/W} \tag{13}$$

Note that $C_\ell/(2a/W)$ is the secant slope to each experimental value of the compliance in Figure 12. As shown in Figure 13, the experimental and analytical correlation is good for crack lengths longer than 2.5 mm (0.1 in.). For smaller crack lengths (1.25 mm and 2.5 mm) the scatter in the experimental results is very high. Any experimental error in compliance measurements causes a large error in the slope $C_\ell/(2a/W)$ for small cracks. Also, the analytical equation for a sharp crack will no longer apply to the real cracks of smaller length as the crack width-to-length

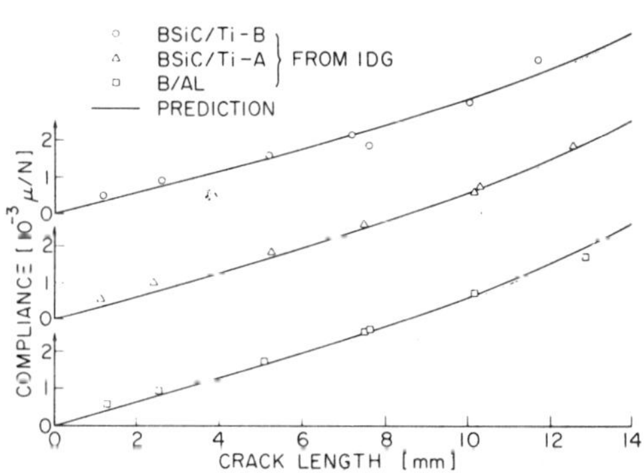

Figure 12. Local compliance vs. crack length.

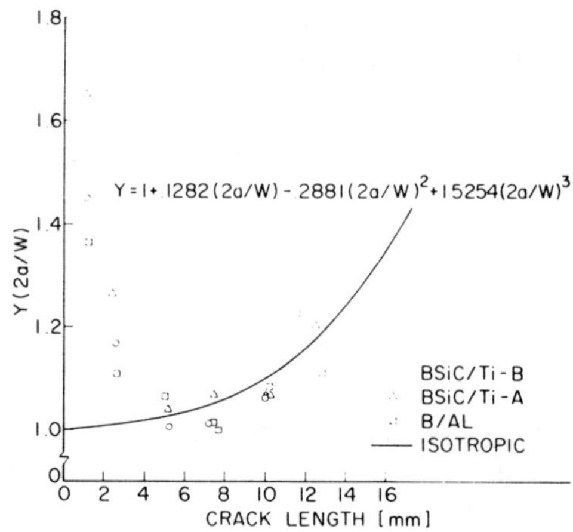

Figure 13. *Comparison of experimental K-calibration factors with theoretical isotropic curve.*

ratio becomes larger. In addition, any crack tip damage will have more significant effect on the elastic compliance so that the initial slope of the subsequent loading-unloading curves may not accurately represent the compliance. However, analytically, the deviation of Y from the isotropic values is negligible as the crack length gets smaller [12], let alone that no K-calibration is needed for small cracks; Y is smaller than 1.001 for cracks having $2a/W$ less than 0.1. Thus, a poor correlation for small cracks is of little significance.

ACKNOWLEDGMENTS

The work of the first author (J. Awerbuch), performed at the Air Force Materials Laboratory (AFML), Wright-Patterson AFB, Ohio, was supported in part by the National Research Council through the Resident Research Associateship Program and in part by the Metals and Ceramic Division of AFML through a contract No. F33615-76-C-5124 with the University of Dayton Research Institute.

The authors wish to express sincere gratitude to Messrs. R. Klinger and C. Bell for their help in the experimental program. Special thanks go to Mr. D. E. Macha for his guidance in the use of the laser interferometric test method and to Dr. N. E. Ashbough for helpful discussions.

REFERENCES

1. T. A. Cruse and J. R. Osias, "Exploratory Development on Fracture Mechanics of Composite Materials," AFML-TR-74-111 Air Force Materials Laboratory, April, 1974.
2. J. W. Mar, "Fracture and Fatigue of Bi-Materials," Mechanics of Composite Review, Air Force Materials Laboratory, Dayton, Ohio, Oct. 1976, p. 117.

3. J. Awerbuch and H. T. Hahn, "Hard Object Impact Damage of Metal Matrix Composites," *J. Composite Materials,* Vol. 10, 1976, p. 231.
4. F. A. Jenkins and H. E. White, *Foundamentals of Optics,* McGraw-Hill, New York, 1957.
5. D. E. Macha, W. N. Sharpe Jr., and A. F. Grant Jr., "A Laser Interferometry Method for Experimental Stress Intensity Factor Calibration," *Cracks and Fracture* ASTM STP 601, American Society for Testing and Materials, 1976, p. 490.
6. W. N. Sharpe Jr., "Interferometric Surface Strain Measurement," *International Journal of Nondestructive Testing,* Vol. 3, 1971, p. 56.
7. W. F. Brown and T. E. Srawley, *Plane Strain Crack Toughness Testing of High Strength Metallic Materials,* ASTM STP 410, American Society for Testing and Materials, 1966.
8. J. Awerbuch and H. T. Hahn, "Crack Tip Damage and Fracture Toughness of Boron/Aluminum Composite," to be published.
9. J. Awerbuch and H. T. Hahn, "Crack Tip Damage and Fracture Toughness of Borsic/Titanium Composite," to be published.
10. J. P. Waszcak, "Applicability of Linear Elastic Fracture Mechanics to 5.6-mil Boron/6061 Aluminum," *J. Aircraft,* Vol. 13, No. 10, 1976, p. 770.
11. D. H. Morris and H. T. Hahn, "Fracture Resistance Characterization of Graphite/Epoxy Composites," *Composite Materials: Testing and Design (4th Conference),* ASTM STP 617, American Society for Testing and Materials, 1977, p. 5.
12. I. M. Daniel, R. E. Rowlands and J. B. Whiteside, "Deformation and Failure of Boron/Epoxy Plate with Circular Holes," *The Test Methods for High Modulus Fibers and Composites,* ASTM STP 521, American Society for Testing and Materials, 1973, p. 143.
13. P. C. Paris and G. C. Sih, "Stress Analysis of Cracks," *Fracture Toughness Testing,* ASTM STP 381, American Society for Testing and Materials, 1965, p. 30.
14. G. C. Sih and H. Liebowitz, "Mathematical Theories of Brittle Fracture" in *Fracture*, Vol. 2, ed. H. Liebowitz, Academic Press, N.Y., 1968.

Fracture Mechanical Investigations on Unidirectional Boron-Aluminum and Boron-Epoxy Composites

PIET PETERS

DFVLR
Institute fur Werkstoff-Forschung
Linder Hohe
D-5000 Koln 90, West Germany

(Received March 28, 1978)

ABSTRACT

The paper presents the results of investigations on DEN (double edge notched) specimens of different crack lengths. With the moire method the shear deformation at the notch tip has been measured. The boron-epoxy composite shows severe shear deformation resulting in matrix cracking parallel to the fibers. Extensive matrix cracking does not occur in notched B-Al. The net fracture strength at crack extension depends on the amount of shear deformation or shear crack formation at the notch tip. This deformation is a function of the crack length. The usefulness of linear elastic fracture mechanics (LEFM) has been investigated with the aid of the net fracture strength at crack extension as a function of the crack length.

INTRODUCTION

MANY INVESTIGATORS HAVE studied the possibility of applying linear elastic fracture mechanics (LEFM) on unidirectional B-Al. According to LEFM the resistance to fracture can be expressed by a critical stress intensity factor at the crack tip, K_c (at unstable crack extension) or K_o (at crack extension) with:

$$K_c = \sigma_c \sqrt{\pi a}\, f(a/W) \tag{1}$$

where:

σ_c = critical stress
a = crack length
$f(a/W)$ = correction factor for the finite width of the specimen.

For the most part the dependence of the K_c (K_o) factor with the crack length has been investigated. This factor should be independent of the crack length, if it represents a material property. Using the isotropic finite width correction, St. John [1] found a dependency of the critical stress intensity factor K_c on the crack length. Iannuzzi [2] also found a rising critical stress intensity factor (K_o) using CN (central notched) specimens, after the validity of isotropic finite width correction was demonstrated.

The compliance measurements of St. John [1] proved to be irreproducable, while Peters [3] found for DEN specimens that the finite width correction could not be described with the usual isotropic width correction. In the latter case the dependence of the factor K_o with the crack length $2a/W$ could be diminished by using the real width correction produced by the compliance measurements. These investigations show that the LEFM is questionable even though Hoover [4] concluded that the LEFM is useful for unidirectional B-Al after finding a toughness (K_o) which is approximately independent of the crack length.

It is clear that when, starting from the notch tip, matrix failure parallel to the fiber direction occurs the fracture toughness cannot be determined with the usual LEFM. This tendency for crack blunting by matrix cracking is always present due to the weak aluminum matrix, compared with the strong boron fibers. In the fracture behaviour the deformation at the crack tip plays an important role. Kreider *et al.* [5] calculated for an elliptical crack in unidirectional B(SiC)-Al that of the possible fracture events shear yielding first occurs, which results in crack blunting. Olster *et al.* [6] recognized the importance of the shear deformation. After estimat ing the length of the plastic zone, they calculated the toughness as a function of the crack length; this showed a good correlation with the experimentally determined curve, based on the LEFM.

Subject of this investigation is again the shear deformation at the crack tip. A relation between the shear deformation and the net fracture strength is sought.

MATERIALS

Boron-epoxy and two different boron-aluminum composites are investigated.

Composite	*Fiber* ϕ	*Fiber volume* %	*Matrix*	*Plate thickness (no. of fibers)*
B-Ep (boron-epoxy)	B 140 μm	53	DX-210 Epoxy	5
B(SiC)-Al	B(SiC) 140 μm	56	6061/4545 Aluminum	5
$B(B_4C)$-Al	$B(B_4C)$ 100 μm	43	6061 Aluminum	5

The boron fibers in aluminum have a SiC or B_4C protection layer.

The B-Ep and $B(B_4C)$-Al composites are produced by DFVLR Stuttgart, the latter in a diffusion bonding process (630°C at 50 atm during 30 min.).

Messerschmitt Boelkow Blohm in Munich has produced the B(SiC)-Al composite in a braze bonding process (600°C at 10^{-2} Torr and 10 atm during 10 min.).

EXPERIMENTS

The fracture behaviour of the composites are investigated on 10 mm wide DEN specimens. The notches (until 3 mm) are made with a diamond wheel (thickness 0.2 mm). The notch radius was 0.10–0.15 mm. The shear deformation is measured with the moire method. Two grilles (on kodalith transparent stripping film), one glued on the specimen and a reference grille glued on a glass-mask with a different number of lines per mm, form in contact fringes. When during loading a shear deformation takes place the originally horizontal fringes (perpendicular to the load) displace. Effectively a reference grille with 33 lines/mm, (fabricated by Delft University of Technology, Department of Aerospace Engineering, Netherlands) and a model grid (1000 points/inch) were used. Instead of a model grille a model grid was used, but apart from the brightness of the fringes this has no effect on the measurements. The moire fringes formed with this grid and grille have a pitch of 0.157 mm. A displacement in loading direction at the notch tip of a pitch-length of the model grid (0.025 mm) results in a fringe displacement of a pitch-length = 0.157 mm; a multiplication factor of 6.

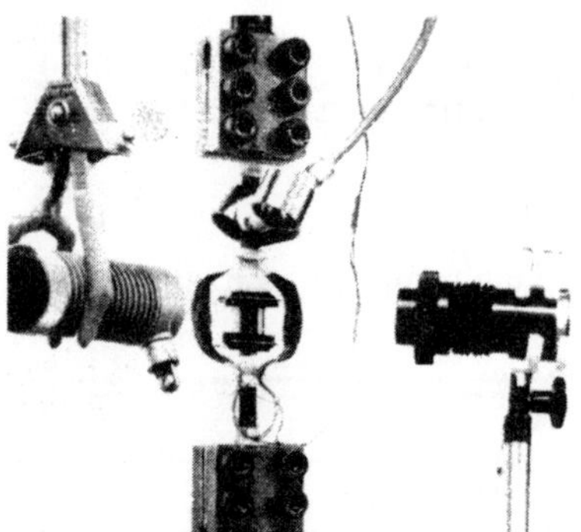

Figure 1. Experimental setup for toughness measurements with piezoelectric acoustic emission transducer, compliance gage, moire masks, and lamps.

The strain as a function of the load is measured with a compliance gage (Figure 1). During the continually running test pictures of the moire pattern have been made.

Crack extension is visible through a sudden displacement increase in the load-displacement curve. In addition, for the B-Ep specimens, some tensile tests on notched B-Ep specimens were run with a piezoelectric acoustic emission transducer; this was necessary in order to be able to detect crack extension as the above method failed to do so.

The fracture behaviour of the composite is dependent on the shear deformation at the crack tip. For this reason the shear strength is an important parameter. The intralaminar shear strength (in the plane of the notch front and the load direction) is estimated with a symmetric block shear test, which showed to be a better method for the determination of shear strength (for unidirectional composites) than the short beam shear test [7]. During this test the middle part of a specimen is sheared off parallel to the fibers by pushing it with a punch in a die (see sketch Figure 10).

RESULTS

Tests on DEN specimens of B-Ep

Figure 2 makes clear that there is a considerable shear deformation at the crack tip, which is extending parallel to the unnotched fibers. This shear deformation leads to shear crack formation parallel to the fibers, which results in complete crack blunting. Finally the specimen ruptures outside the original notch plane.

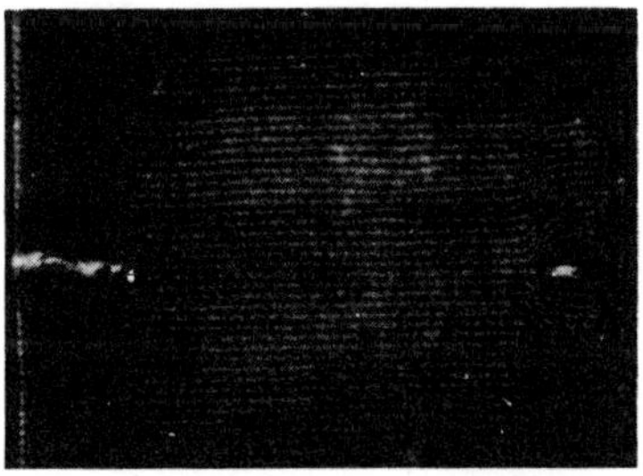

Figure 2. The shear deformation at the crack tip of a boron-epoxy specimen (2a/W=0.29) at 95% of the fracture load.

In Figure 2 the shear deformation is so high, that shear cracks already must have been formed, which however are not visible in the moire pattern as the grid glued on the specimen remains undisturbed. Only for very small crack lengths ($2a/W =$ 0.05–0.1) fracture takes place colinear with the original notch. A single notch length exceeding 3 x the fiber pitch ($2a/W > 0.1$) is sufficient for the formation of shear cracks in the matrix. At increasing crack length shear cracking occurs at a decreasing net stress. The shear cracking, when starting abruptly, causes a sudden elongation with possibly a load drop. The load deflection curve however does not detect fiber fracture. Also with the acoustic emission transducer it was not possible to detect fiber fracture caused by the stress concentration at the crack tip. The stress-sound curve shows at matrix crack formation a sudden sound count-increase (Figure 3). It seems, that if fiber rupture occurs at the notch tip the sound impulse is so small, that it is lost in the other sound producing fracture events.

Because of the fact that the load at crack extension colinear with the original notch could not be determined, only the stress at specimen fracture is presented in Figure 4. This makes clear that the rupture stress is approximately independent of the crack length.

Tests on DEN specimens of B-Al

The specimens of both B-Al composites all broke colinear with the original notch. At the notch tip, during loading a shear deformation develops, which does not lead to extensive matrix cracking parallel to the fibers. At a small notch length crack extension immediately leads to specimen rupture. Only at higher crack length stable crack extension occurs, which for sufficient crack extension is visible in the moire pattern (Figure 5). The shear deformation is measured in the pictures made of the moire pattern at the different loads. The change in position of the fringes just above and below the notches 2½ times the fringe pitch before and after the notch tip has been measured (see sketch Figure 6). This distance has been chosen,

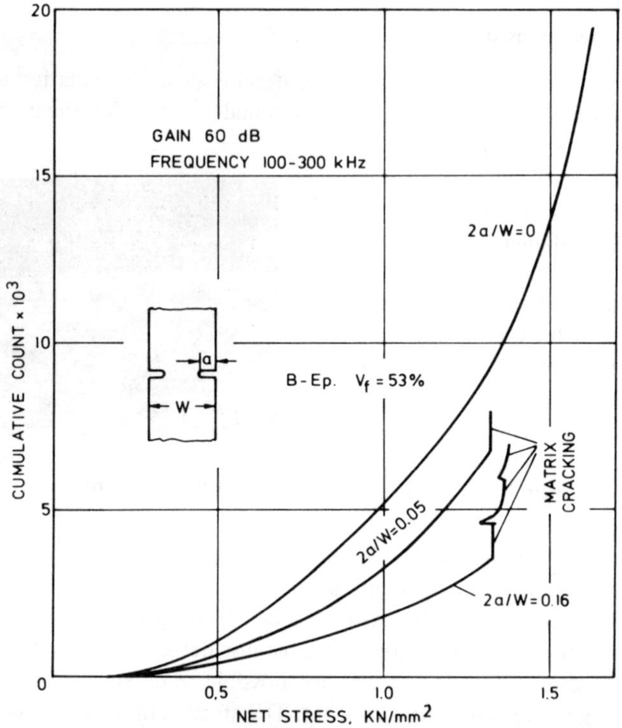

Figure 3. Acoustic emission from fracture tests on DEN boron-epoxy specimens with different crack lengths.

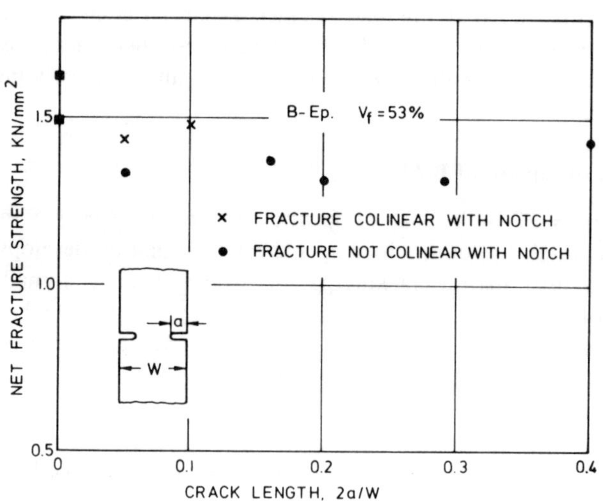

Figure 4. The variation of the net specimen fracture stress with the crack length (2a/W) for boron-epoxy (V_f = 53%).

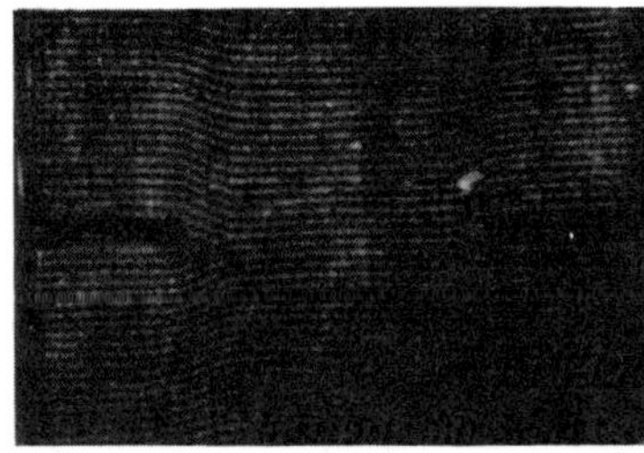

Figure 5. A DEN $B(B_4C)$-Al specimen ($2a/W$ = 0.52) at 97% of the fracture load showing shear deformation and crack extension at both notches.

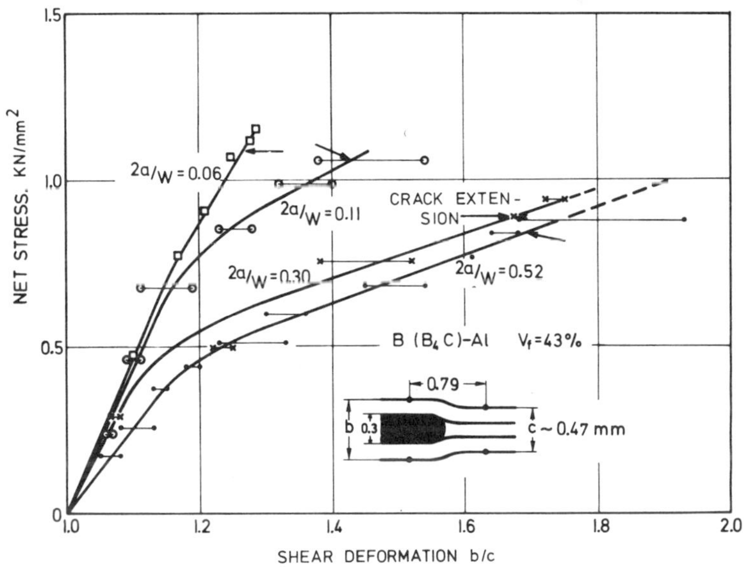

Figure 6. The net stress versus shear deformation b/c (measured in the moire pattern) for DEN $B(B_4C)$-Al specimens with different crack lengths.

because the complete measureable shear deformation at the notch tip takes place in this area. The distances between these fringes after and before the notch tip are divided, which gives a degree of shear deformation.

The shear deformation as a function of the stress for different crack lengths of $B(B_4C)$-Al specimens are presented in Figure 6. The shear deformation at both notch tips are considered, except for the small crack length. In the case of the latter the smaller shear deformation demands a higher magnification, so that only one notch tip could be observed. At this small crack length ($2a/W = 0.06$) the shear deformation at the crack tip is mostly elastic. For greater crack lengths the plastic deformation increases before crack extension and specimen fracture occur. When crack extension disturbs the shear reference length c, the curve is drawn intermittently until the fracture stress (Figure 6, for $2a/W = 0.30$ and 0.52). In Figure 7 the

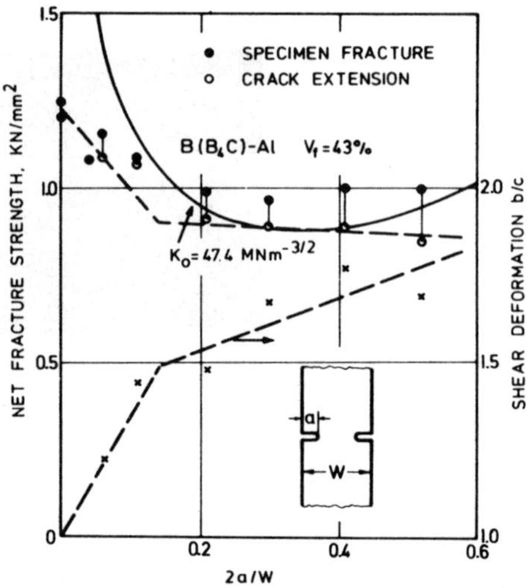

Figure 7. The variation of the net fracture stress at crack extension and specimen rupture and the variation of the shear deformation with crack length (2a/W) for B(B_4C)-Al.

net fracture stress at crack extension (open symbol) and specimen fracture (closed symbol) is presented for B(B_4C)-Al as a function of the crack length $2a/W$. Only at greater crack lengths is there stable crack extension, which, however, is not considered for the calculation of the net fracture stress at specimen rupture. It can be seen that the net fracture stress at crack extension is strongly correlated to the degree of shear deformation at crack extension.

In the range where the shear deformation is mainly elastic, the net fracture stress decreases quickly at increasing crack length until for still greater crack lengths the plastic shear deformation blunts the crack. This results in a less steep decrease of the net fracture stress at increasing crack length (Figure 7).

B(SiC)-Al shows in principle the same effect (Figure 8). Again for a small crack length there is a sharp decrease in the net fracture strength. In the plastic area the notch is blunted and a tendency for a net fracture strength increase with increasing crack length is observed.

Both B-Al materials have about the same tensile strength in the unnotched condition, which shows that the B(SiC) fiber because of the higher fiber volume of B(SiC)-Al has a lower strength than the B(B_4C) fiber. The differences in fracture behaviour between the B(SiC)-Al and B(B_4C)-Al are:

1. In the elastic range the net fracture stress (at crack extension) of B(SiC)-Al decreases faster.

2. B(SiC)-Al reaches a relatively smaller net fracture stress level in the plastic range.

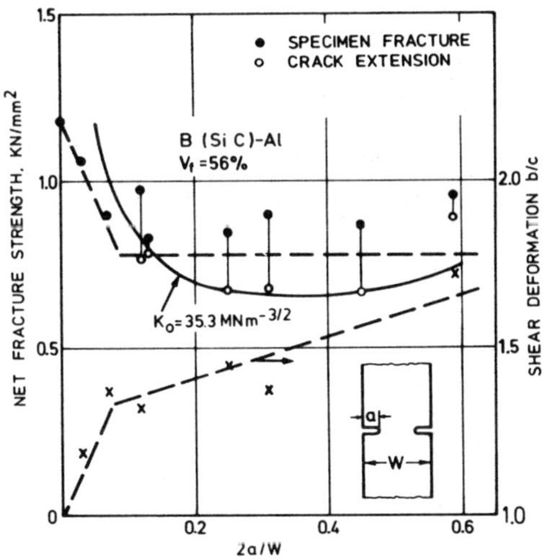

Figure 8. The variation of the net fracture stress at crack extension and specimen rupture and the variation of the shear deformation with crack length (2a/W) for B(SiC)-Al.

3. For crack lengths greater than $2a/W = 0.4$ B(SiC)-Al shows a rising, B(B_4C)-Al a decreasing, net fracture strength with increasing crack length.

Only in one example a small shear crack (~0.5 mm at both notch tips of a B(B_4C)-Al specimen with $2a/W = 0.41$) could be observed in the moire pattern. After fracture some specimens of both B-Al materials were investigated in the scanning electron microscope. Particularly the front of the original notch was examined. These investigations showed, that at increasing notch length near the notch tip an increasing number of cut fibers are split. The split fibers are partly connected by matrix cracks. A few examples are shown in Figure 9 for the B(SiC)-Al composite at different notch lengths. This makes it clear that even though no sign of matrix cracking can be observed at the surface the initial onset of matrix cracking can occur inside the specimen.

The Shear Strength of the Investigated Materials

The three unidirectional composites are investigated in symmetric block shear testing in order to estimate the shear strength and the resistance to matrix cracking. The test technique is evaluated in [7]. In this test the middle part of a specimen is sheared off by pressing it with a punch in a die.

Typical shear stress-deflection curves for the three materials are presented in Figure 10. The maximum stress represents the shear strength, while the area under the curve is a measurement of the shear energy, which is necessary for shear crack formation and propagation.

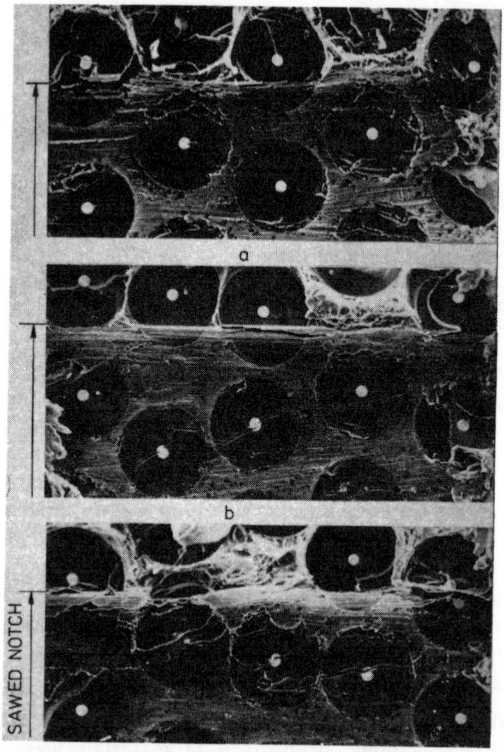

Figure 9. Scanning electron photomicrographs of the notch tip of DEN B(SiC)-Al specimens after fracture. (a) 2a/W = 0.12; (b) 2a/W = 0.30; (c) 2a/W = 0.59.

The mean shear strengths are:

B(SiC)-Al $\tau_B = 109.0\ \text{N/mm}^2$

B(B_4C)-Al $\tau_B = 86.3\ \text{N/mm}^2$

B-Ep $\tau_B = 49.8\ \text{N/mm}^2$

B-Ep shows the lowest shear strength and resistance to matrix cracking, while B(SiC)-Al has the best properties.

DISCUSSION

The investigated materials showed two clearly different fracture phenomena:

1. Matrix cracking parallel to the fibers followed by specimen fracture not colinear with the original notches (B-Ep).

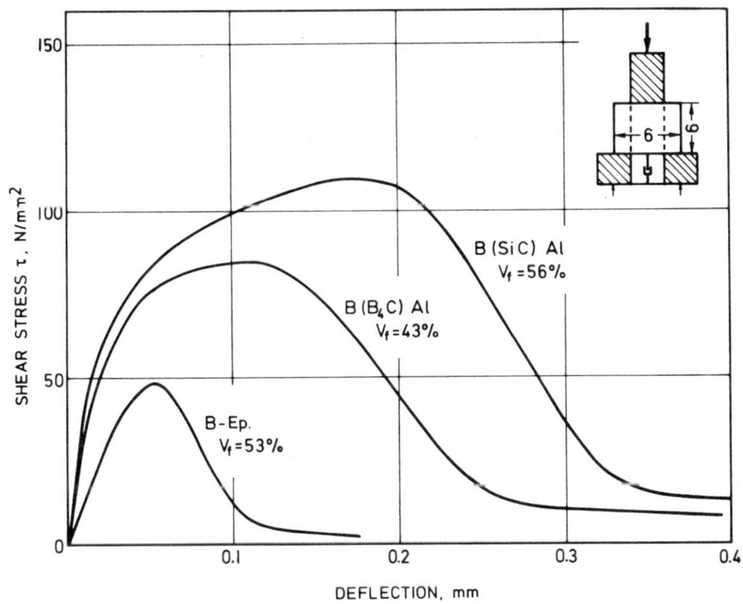

Figure 10. Typical shear stress deflection traces obtained from symmetric block shear tests on the investigated composites.

2. Crack extension and specimen fracture colinear with the notch plane (B-Al).

It is clear that the LEFM cannot describe specimen fracture when through matrix cracking the notch is completely blunted, as is the case for B-Ep. This matrix cracking parallel to the fibers can be understood when the low shear strength and shear deformation (Figure 10) for B-Ep is considered. The shear strengths of both B-Al composites, however, are so high that crack propagation through the fibers at the notch tip occurs instead of matrix cracking parallel to the fibers.

Can the LEFM describe specimen fracture of both the B-Al composites?

To answer this question a relation between the net fracture stress and the fracture toughness K_o (at crack extension) is considered; this is developed through modification of (1). This relation between σ_{net} and K_o is:

$$\sigma_{net} = K_o/(\sqrt{\pi a}\,(1 - 2a/W) f(2a/W)) \tag{2}$$

where $f(2a/W)$ is the width correction factor after Brown and Srawley [8]

$$f(2a/W) = 1.12 + 0.20\,(2a/W) - 1.20\,(2a/W)^2 + 1.93\,(2a/W)^3 \tag{3}$$

Based on the relation (2) for $B(B_4C)$-Al a net fracture strength-crack length curve is drawn in Figure 7 assuming the fracture toughness $K_o = 47.4\ MN/m^{3/2}$.

This K_o value is determined by forcing the curve resulting from equation (2) in Figure 7 through the middle crack length ($0.2 < 2a/W < 0.4$) data. At lower and higher crack lengths the experimental strengths lie below the $K_o = 47.4\ MN/m^{3/2}$ curve, which shows that in reality K_o is not constant but shows a maximum at middle crack lengths.

The B(SiC)-Al composite fits the $K_o = 35.3\ MN/m^{3/2}$ curve better because of the strong drop of the net fracture strength in the elastic area and the tendency of increasing net fracture strength at increasing crack length in the plastic range. One is tempted to recognize the usefulness of the LEFM because of this good agreement. However, it is likely, that the rising net fracture stress in the plastic range is caused by scatter and the onset of matrix cracking; none the less the rising fracture stress can not be caused by a less severe stress concentration at the crack tip, as the shear deformation increases in the plastic area at increasing crack length (Figure 8).

St. John [1] also found a rather steep rising net fracture strength for B-Al 99.4 at higher crack length, which he bases on the ductility of the matrix and matrix-matrix bond failure.

Both materials show for small crack lengths ($2a/W < 0.1$) mainly elastic shear deformation and for larger crack lengths an increasing additional plastic shear deformation. The net fracture stress is strongly correlated to this deformation. The differences of the three materials show that the net fracture stress is strongly influenced by parameters such as fiber strength and stiffness, matrix (shear) strength and module, and the fiber volume. The LEFM does not consider these parameters so that in general the LEFM is not useful. The net fracture stress seems to be a better criterium for fracture.

It is difficult to explain the differences in the net-fracture curves (Figures 7, 8) as there are too many independent variables. The lower strength of the B(SiC) fiber, as observed before, probably causes the steeper drop of the net fracture stress curve in the elastic range, although the somewhat higher shear deformation in this range could contribute to this effect. In the plastic range crack extension in B(SiC)-Al takes place at a smaller shear deformation than in $B(B_4C)$-Al. In spite of smaller shear deformation the net fracture stress at crack extension is lower; this can be explained with the lower strength of the B(SiC) fiber.

A better understanding of the net fracture stress-crack length relation demands an examination of the influence of the unique composite parameters such as matrix shear strength and stiffness, fiber strength and stiffness, and the fiber volume.

These shear deformation measurements cannot directly result in a shear concentration factor because the shear deformation distribution in the investigated area is not examined. Another disadvantage is that on a very small scale, where shear cracking begins, the moire technique based on glued films is not very exact (as Figure 2 shows).

CONCLUSION

This investigation of unidirectional fiber reinforced materials shows that composites with a low shear strength (B-Ep) show complete crack blunting by the formation of matrix cracks parallel to the fibers. This results in a net fracture strength independent of the crack length. The composites with a higher shear strength (B(SiC)-Al, $B(B_4C)$-Al) show fracture colinear with the notches. The net fracture stress (at crack extension) is strongly related to the shear deformation at the notch tip. At small notches ($2a/W < 0.1$) there is mainly an elastic deformation which causes a sharp drop in the net fracture stress (Figure 7, 8). At larger crack lengths the matrix shear yielding and onset of shear cracks blunt the notches, which results for the $B(B_4C)$-Al composite ($V_f = 43\%$) in a less sharp drop of the net fracture strength at increasing notch length. The B(SiC)-Al composite ($V_f = 56\%$) shows in the plastic range a small increase in net fracture stress; This is possibly caused by scatter and shear crack formation. The relation net fracture stress to crack length of the LEFM based on a constant fracture toughness (K_o) shows a minimum at a middle crack length (Figure 7, 8). The test results on B(SiC)-Al specimens fit the K_o = constant curve best. Dependent on the matrix shear strength and modulus, fiber strength and stiffness, and fiber volume of the composite, a wide variety of net fracture strength/crack length relations are possible. Because of the fact that the LEFM does not account for these variables, the LEFM is in general not useful, even when the crack extends colinear with the original notch.

REFERENCES

1. C. F. St. John and K. N. Street, 3. International Conference on Fracture Mechanics, Munich 1973, VIII 333.
2. F. Iannuzzi, Ph.D. Thesis, The University of Tennessee, 1974.
3. P. W. M. Peters, and D. Munz, Anwendung der linear elastischen Bruchmechanik auf das Bruchverhalten von B/Al Verbundwerkstoffen, Conference of the Deutsche Gesellschaft für Metallkunde, Konstanz, April 1975.
4. W. R. Hoover, "Crack Initiation in B-Al Composites," *J. Comp. Mat.* Vol. 10 (1976), p. 106.
5. K. G. Kreider, L. Dardi, "Fracture Toughness of Composites," Report L 110865-1, United Aircraft Research Laboratories, 1972.
6. E. F. Olster, R. C. Jones, "Toughening Mechanisms in Continuous Filament unidirectionally Reinforced Composites," ASTM STP 497, 1972, p. 189.
7. P. W. M. Peters, "The interlaminar shear strength of B-Al Composites," *J. Comp. Mat.* Vol. 12 (1978), p. 53.
8. W. F. Brown and J. E. Srawley, ASTM STP 410 (1966).

Crack-Tip Damage and Fracture Toughness of Boron/Aluminum Composites

JONATHAN AWERBUCH

Department of Mechanical Engineering and Mechanics
College of Engineering
Drexel University
Philadelphia, PA 19104

AND

H. THOMAS HAHN*

Nonmetallic Materials Division
Air Force Materials Laboratory
Wright Patterson AFB, OH 45433

(Received January 29, 1979)

ABSTRACT

The fracture behavior of center notched unidirectional boron/aluminum composite under tensile loading has been investigated. Load-crack opening displacement (COD) curves were obtained for crack length-to-width ratios varying from 0.05 to 0.5. The COD was measured across the crack surfaces by means of the laser interferometric technique (gage length of 500 μm). As expected, the interferometric displacement gage (IDG) was found to be very sensitive to the appearance of damage at the crack tip, resulting in highly nonlinear load-COD curves. The damage at the crack-tip has been carefully examined by radiographs, SEM, photomicrographs, interferometric miscroscope and optical interferometry technique. Fracture strength results and experimental load-COD curves were compared with analytical predictions and an excellent agreement has been established. The resistance curve method has also been tried and found to be non-applicable. An understanding of the correlation between the observed failure modes and deformation characteristics has been achieved.

*Presently, University of California, Lawrence Livermore Laboratory, Livermore, CA 94550.

INTRODUCTION

BORON/ALUMINUM COMPOSITES have received most of the attention, among a variety of metal matrix composites, in the research activity and applications studies of recent years [1]. The mechanical behavior has been studied extensively during the past decade and results published in many reports and publications. An important subject of concern is the fracture behavior of boron/aluminum composites, which exhibit extremely low elongation up to fracture. The importance of understanding the fracture behavior of boron/aluminum can be demonstrated, for example, from the results of recent study on hard object impact damage of metal matrix composites [2]; the induced impact damage can be modeled as through-the-thickness cracks for post-impact residual strength predictions. In other words, the fracture behavior of unidirectional boron/aluminum composite with induced lateral impact damage is similar to the fracture behavior in the presence of artificial cracks and holes. Thus, understanding the behavior of the material in the presence of artificial notches is essential for design purposes.

A fair amount of work on the fracture behavior and toughness of boron/aluminum (B/Al) composites has been performed in recent years. The applicability of linear elastic fracture mechanics (LEFM) to unidirectional B/Al has been discussed in [3, 4, 5] with varying conclusions. For example, Kreider and Dardi [3], and Mar [4] concluded that the correlation between fracture strength σ_f and half crack length a is best described by an equation of the form $\sigma_f a^m$ = Const. However, contrary to LEFM predictions, the exponent m was found to be less than one half. Wright [5] concluded that LEFM can be applied, provided that the crack extension prior to failure is added to the initial crack length. With this assumption, the data of [5] as well as [3] do agree with predictions based on LEFM. This assumption does follow, in fact, the analytical models proposed in [6, 7] for notched strength predictions of composites. Hancock and Swanson [8] concluded that the conventional concepts of fracture toughness can be useful in characterizing the toughness of unidirectional B/Al. Sun and Prewo [9] showed that the fracture toughness of B/Al can be described by LEFM provided the crack extension is colinear with the initial pre-machined notch, as is the case of [90] or [0/90] laminate of compact tension specimen geometry. For unidirectional B/Al, such an approach is not relevant [9] due to the large amount of plastic deformation of the matrix parallel to the filaments. Similar damage growth and modes of failure under static and fatigue loadings have been observed and discussed in detail by many authors, e.g. [3, 10, 11, 12]. In general, it has been shown that the failure of unidirectional B/Al is preceded by longitudinal plastic deformation at the crack tip which eventually leads to matrix cracking and crack-tip blunting, all of which contribute to the crack arresting mechanism. The incipient matrix cracking has been described by the elastic stress intensity factor [10]. Obviously when different analytical approaches and experimental techniques are employed to define fracture toughness, the resulting toughness values vary significantly. However, most of the values are still very promising in comparison to other metal or resin matrix composites. Although many studies have been performed on the fracture behavior of B/Al composites, a definitive correlation between the observed failure modes and the deformation characteristics is lacking at this stage. Therefore, in the present paper, more emphasis is placed on the deformation behavior and the associated characterization methods.

Specifically, load-crack opening displacement (COD) curves are presented for center notch unidirectional 5.6 mil B/Al-6061-F specimens having crack length-to-width ratios ranging from 0.05 to 0.5. The COD was measured across the crack surface by means of laser interferometric technique. As expected, the Interferometric Displacement Gage (IDG) is found to be very sensitive to the appearance of crack tip damage, resulting in highly nonlinear load-COD curves.

The damage at the crack tip is examined by radiography, SEM, photomicrographs, interferometric microscope, and optical interferometry technique. Fracture strength data are analyzed and load-COD curves are compared with predictions, and an excellent agreement between theory and experiments is established. The resistance curve method was also tried to characterize the fracture resistance of the subject material but could not be applied. The results of this work, in conjunction with the results of a recent study [13] on the K-calibration of unidirectional metal matrix composites, gives a detailed description of the fracture behavior of unidirectional boron/aluminum.

EXPERIMENTAL PROCEDURE

Displacement Gages

The crack opening displacement (COD) was measured by means of laser interferometric technique which enables measuring displacement over a very short gage length. The COD was measured across the crack surface, at the center of the crack, with gage length of about 500 μm (0.02 in). This Interferometric Displacement Gage (IDG) was found to be highly sensitive in detecting crack-tip damage, very accurate in the submicron scale, and simple to utilize. As the Interferometric Displacement Gage (IDG) has not been used before for composite materials, it was also of interest to compare the load-COD curves with the load-displacement curves obtained by using standard compliance and clip gages, 1.0 and 0.25 inch gage length, respectively. A detailed description of the test procedure and experimental technique of both the IDG and clip gages is given in [13]. A comprehensive description of the laser interferometric technique is outlined in several papers published recently, e.g. [14, 15]. All tests were performed on the Instron testing machine at displacement rate of 0.005 inch/min.

Materials

The material tested is unidirectional boron/aluminum 6061-F (B/Al-6061-F) containing approximately 50% by volume 5.6 mil diameter boron filaments. The material was fabricated by United Technology Research Center by the "Quick Bond" process using the conditions of 550°C, 68.9 MPa, 10 min. in air. All specimens were cut from 5.5 X 5.5 inch plates, 0.052 inch thick (eight plies), parallel to the fibers' direction with a diamond saw or by E.D.M. (Electro Discharge Machining) method, and were ground on the edges to ensure minimum damage. The specimen dimensions were 14cm. long by 2.5cm. wide (5.5 in. by 1.0 in.). Center notches of about 0.05, 0.1, 0.2, 0.3, 0.4, 0.5 in. long and 0.01 in. wide were introduced by E.D.M. Aluminum doublers of 1.0 in. square were sheared from 0.05 in. thick 2024-T6 aluminum sheets and were glued with epoxy onto the specimen ends. A total of 16 center notch tension specimens were tested. Several unnotched specimens were tested statically to provide baseline strength data.

Failure Modes

A detailed study of the crack-tip damage growth and fracture surface has been performed by radiography, SEM, photomicrographs, interferometric microscope, and optical interferometry technique. Each specimen was loaded several times to a predetermined load, unloaded, and removed from the testing machine. Through photomicrographs the crack tip damage was carefully examined at the outer surface of the specimen, and the permanent opening of the crack was measured. The damage growth resulting from increasing load was also examined using the optical interferometry technique. The fracture surface was inspected through the SEM. For a few specimens the matrix was dissolved and the damage extension beyond the fracture surface was examined.

RESULTS AND ANALYSIS

Load-COD Curves

A typical load-COD curve obtained from the Interferometric Displacement Gage (IDG) is shown in Figure 1 for a B/Al specimen having initial crack length of 12.85 mm (0.504 in.). As seen in Figures 1 and 2, the load-COD curve is highly nonlinear, followed by rapid increase in COD when the load approaches its ultimate level. The rapid changes in COD have been found to be due to fiber failure, longitudinal plastic deformation, and longitudinal cracking of the matrix near the crack tip. This type of damage and plastic deformation at the crack tip is thus easily detected by monitoring the COD at the center of the crack.

The short gage length (500μm) of the IDG makes this technique more sensitive to the appearance of crack tip damage than the standard compliance gages, clip gages, or strain gages located at the crack tip. In fact, all previous load displacement curves obtained for unidirectional B/Al and other metal matrix and resin matrix composites are linear, as presented in numerous publications [16, 17, 18]. When the IDG is used, the crack tip damage is manifested in highly nonlinear load-COD

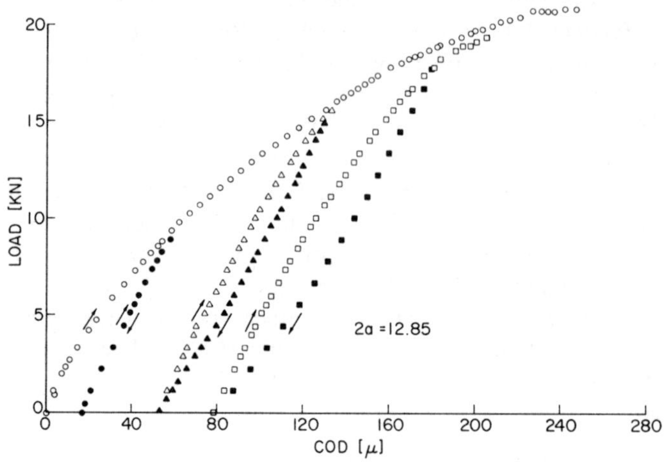

Figure 1. Load-COD (loading-unloading) curves obtained with interferometric displacement gage (crack length 12.85mm). The high nonlinearity and the large permanent COD remaining upon unloading indicate severe crack tip damage.

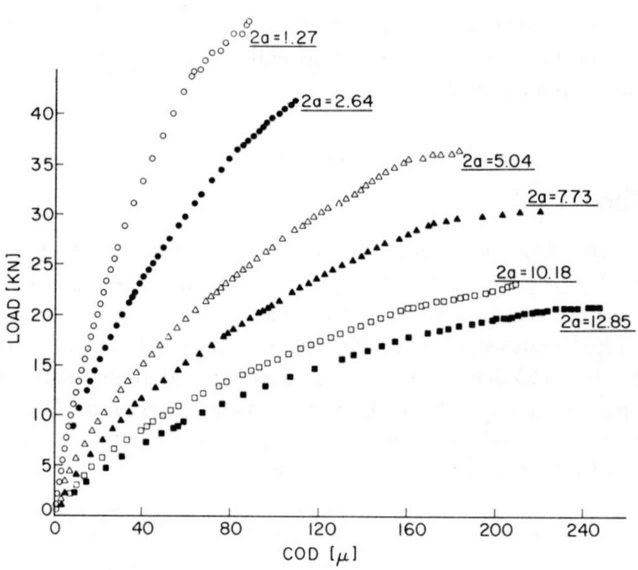

Figure 2. Load-COD curves for various crack sizes.

curves and a large permanent COD remaining upon unloading, as shown in Figure 1. To verify the accuracy of the large permanent displacement obtained with IDG technique, each specimen was removed from the testing machine after each unloading and the permanent opening of the crack was directly measured under a microscope. A comparison of the permanent displacement measured from the load-COD curves and from the direct measurements shows an excellent agreement for all specimens and for all load levels, demonstrating the accuracy of the IDG technique.

As the load approaches its ultimate level, the crack tip damage grows much more rapidly, causing the rapid increase in COD. Detailed examination of the load-COD curves also reveals a high nonlinearity between jumps in COD.

This high nonlinearity of the load-COD curve makes it difficult, and practically impossible, to measure the elastic compliance, and consequently to determine the stress intensity factor calibration curve. Therefore, in using the IDG technique, each specimen was loaded to a predetermined load and then unloaded. This procedure was repeated several times incrementally up to failure, as shown in Figure 1. Examining the load-COD curve (Figure 1) reveals that the unloading curve and the subsequent loading curve are initially linear. It is this linear portion of the load-COD curve that was used in [13] to obtain the compliance as a function of the initial crack length. Although the nonlinearity of the load-COD curves and the large permanent COD obtained with the IDG indicate the occurance of crack tip damage at relatively low load levels, this damage is subcritical and does not affect the compliance measurements. A detailed discussion of the compliance curves and the stress intensity factor calibration is given in [13].

Figure 1 also shows a "combined" load-COD curve, constructed by combining the individual loading-unloading curves. As seen in Figure 1, the overall load-COD behavior is very similar to that of metals. The effect of the crack length on the load-COD curve is shown in Figure 2. Here again the "combined" load-COD curves are shown for each crack length. Although the details of each curve cannot be seen in Figure 2, the nonlinearity, rapid increase in COD, and permanent opening upon unloading were obtained for every crack length tested. As expected, these characteristics are more pronounced as crack length increases; that is, the longer the original crack length, the larger the crack tip damage.

It may be noted that damage growth is accompanied by audible levels of acoustic emission. A recent study correlating the acoustic emission with damage growth detected through COD measurement for B/Al-1100 shows that a qualitative correlation can be found [19].

Fracture Strength

Table 1 lists the fracture strength data for various crack lengths. The average unnotched strength (of seven specimens) was relatively high; 2004 MPa (290.7 ksi) with standard deviation of 82.7 MPa (12.0 ksi). The average longitudinal stiffness was 245.5 GPa (35.6×10^3 ksi) with standard deviation of 14.8 GPa (2.15×10^3 ksi).

Table 1: Notched Strength of Unidirectional Boron/Aluminum

Crack Length [mm]	Crack Length [inch]	Notched Strength [MPa]	Notched Strength [ksi]
1.27	0.050	1537.7	222.30
1.35	0.053	1493.6	216.62
2.64	0.104	1282.1	185.95
2.64	0.104	1321.9	191.73
5.03	0.198	1112.1	161.30
5.23	0.206	1089.6	158.03
7.47	0.294	896.6	130.04
7.59	0.299	897.4	130.15
7.72	0.304	954.0	138.37
10.18	0.401	708.1	102.70
10.18	0.401	719.4	104.34
12.85	0.506	625.2	90.68
12.85	0.506	602.2	87.34

The notched strength of composite materials can be predicted by using the existing analytical model suggested by Waddoups, Eisenmann and Kaminski (WEK), [6]. They assumed the existence of intense energy regions at the edges of the notch which are modeled as edge cracks of size c_o. With this in mind, the linear elastic fracture mechanics is applied and the fracture strength can be predicted according to:

$$Y\frac{\sigma_f}{\sigma_o} = \left[\frac{c_o}{a + c_o}\right]^{1/2} \tag{1}$$

where σ_f and σ_o are the notched and unnotched strengths, respectively. $Y(2a/W)$ is the isotropic finite width correction factor and a is the half crack length.

For the finite width correction factor Y, the following expression has been chosen [20]:

$$Y = 1 + 0.1282\,(2a/W) - 0.2881\,(2a/W)^2 + 1.5254\,(2a/W)^3 \tag{2}$$

Note that the analytical prediction uses the isotropic width correction factor, Equation (2), because of its proven applicability to unidirectinal B/Al specimens of the geometry chosen in the present work [13].

A similar model has been mathematically formulated by Whitney and Nuismer (WN), [6, 7]. Here the assumption is that failure occurs when the average stress over a distance a_o, ahead of the crack tip, equals the unnotched strength. From this "average stress criterion" the fracture strength can be predicted as

$$Y\frac{\sigma_f}{\sigma_o} = \left[\frac{a_o/2}{a + a_o/2}\right]^{1/2} \tag{3}$$

By comparing both models (Equations (1) and (3)) it can be seen that they are identical ($a_o = 2c_o$). It should be noted that c_o (or a_o) can only be obtained experimentally by performing at least one fracture test, and must be determined for each composite system independently. Since c_o (or a_o) is assumed to be a material property, i.e. a quantity independent of crack size, rearranging Equation (1) will give c_o as the slope of the following equation:

$$a = c_o\left[\left(\frac{\sigma_o}{Y\sigma_f}\right)^2 - 1\right] \tag{4}$$

Figure 3 shows the results plotted according to the format of Equation (4). The solid line represents a least squares fit of c_o which was found to be 1.0mm (0.0396 in.). As seen in Figure 3, the use of a constant value for c_o for different crack lengths appears to be justified. Thus, the effect of crack size on notched strength can be described very well by Equation (1) with c_o = 1.0mm (Figure 4). Note that a significant notch sensitivity is observed for the unidirectional B/Al; the notched strength at $2a/W = 0.5$ is only 40% of the unnotched strength.

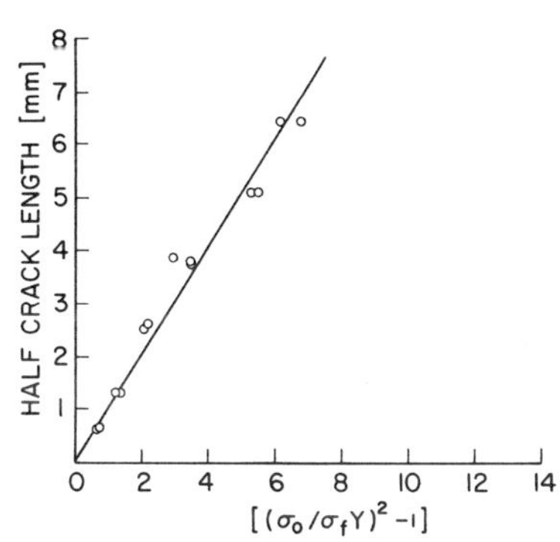

Figure 3. Least squares fit for c_o.

Results obtained from [21] for the fracture strength data of unidirectional B/Al-6061 were analyzed as well. These B/Al composites were fabricated by Amercom Inc. through diffusion bonding, employing three different combinations of diffusion temperatures, pressures, and time. The details of fabrication procedure were not accessible to us, being the company proprietary information. Figure 5 shows the results of c_o for the three different materials. Again, the solid lines represent a least squares fit of c_o, and, as before, it seems that a constant value can be used. However, Figure 5 shows that each material has a different value of c_o. The numerical values of c_o for the three different B/Al are listed in Table 2.

As seen in Equation (1), smaller values of c_o indicate higher notch sensitivity, i.e., steeper change of $Y\sigma_f/\sigma_o$ with a. Thus, the conclusion should be that the fabrication procedure (pressure, temperature, and time of diffusion bonding

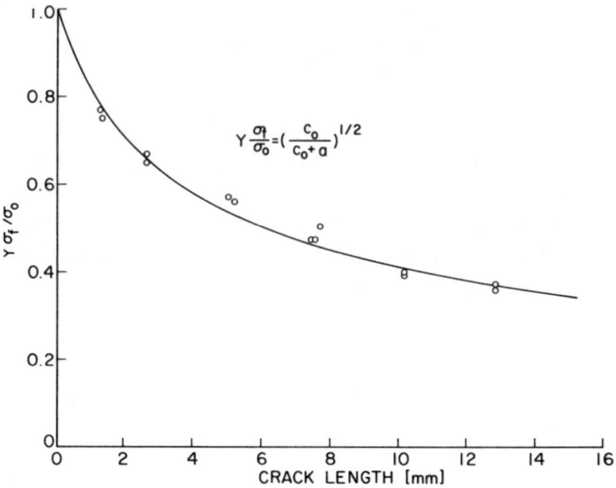

Figure 4. Comparison of analytical (WEK model) and experimental results for notched strength. Results show large notch sensitivity.

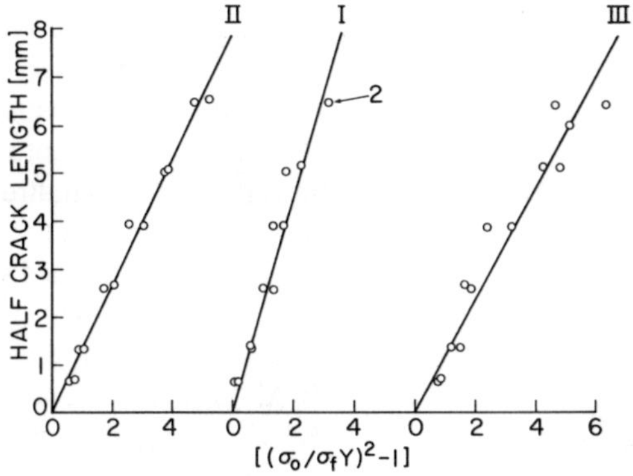

Figure 5. Least squares fit for c_o for various unidirectional B/Al composites. The fabrication procedure affects c_o.

procedure) does affect to some extent the notch sensitivity of B/Al composites. Values of the critical stress intensity factor, K_c obtained from the equation:

$$K_c = Y\sigma_f \sqrt{\pi(a + c_o)} = \sigma_o \sqrt{\pi c_o} \tag{5}$$

are also given in Table 2, indicating that the notch sensitivity depends, in fact, on the fabrication procedure. Figure 6 shows both experimental results and analytical

Table 2. Notched Strength Data

	σ_o			K_c	c_o
Material	*Average MPa (ksi)*	*S.D.[a] MPa (ksi)*	*C.V.[b] %*	*MPa$\sqrt{m}$ (ksi$\sqrt{in}$)*	*mm (inch)*
B/Al-present	2004 (290.7)	82.7 (12.0)	4.11	112.7 (102.6)	1.00 (0.0396)
B/Al-Type I	1290 (187.1)	60.1 (8.7)	4.67	107.3 (97.6)	1.15 (0.0455)
B/Al-Type II	1432 (207.7)	88.1 (12.8)	6.16	91.8 (83.5)	1.31 (0.0515)
B/Al-Type III	1526 (221.3)	65.7 (9.5)	4.31	91.9 (83.7)	2.20 (0.0867)

[a]Standard Deviation
[b]Coefficient of Variation

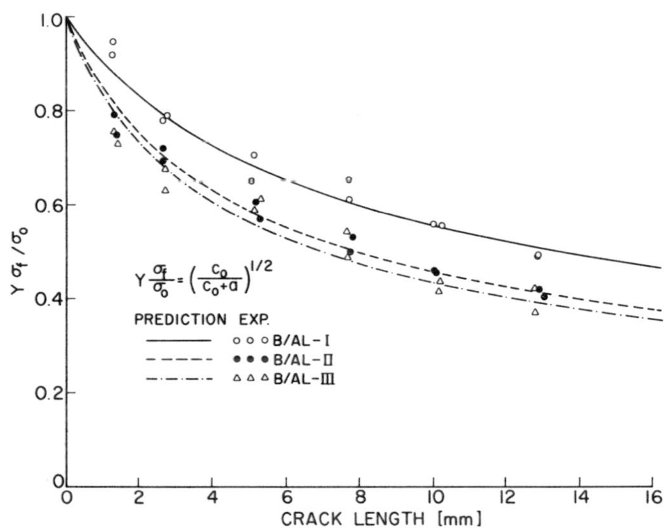

***Figure 6.** Effect of fabrication procedure on notch sensitivity. Best-fit curves (WEK model) are also shown.*

predictions, Equation (1), for the three different B/Al systems, based on c_o values of Table 2. The predictions represent the experimental results very well, except for the very short crack length of 0.05 in. That is, for B/Al type I, the model predicts higher strength than is actually observed, while the opposite holds true for B/Al types II and III.

As mentioned briefly in the introduction, another approach suggested by Mar and Lin (ML-model) [4] to analyze the fracture strength data is to assume

$$\sigma_f = M(2a)^{-m} \tag{6}$$

where M and m are determined so as to fit the data best. This model interprets m to be "the order of singularity of a crack lying in the matrix with its tip at the interface of matrix and filament." It was found in [4] that Equation (6) closely fits the data for $[\pm 45/0_2]$ B/Al, $[0/\pm 45]_{2s}$ graphite/epoxy [6], and $[0/\pm 45/90]_{2s}$ graphite/epoxy [7], and others. A similar model was presented in [22] and was supported experimentally in [3].

The data obtained in this program are plotted on log σ_f/σ_o versus log $2a/W$ format (Figure 7). A least squares fit yields the values of the slope m and of the coefficient M as 0.321 and 0.313, respectively, while the coefficient of correlation (c.c) is 0.971. As seen in Figure 7, Equation (6) follows the data quite closely. However, in order to compare both approaches, Equations (1) and (6), the WEK model is also plotted on the logarithmic format in Figure 8. A comparison of both Figures 7 and 8 reveals that the first model, Equation (1), does represent more accurately the notch sensitivity of unidirectional B/Al composites.

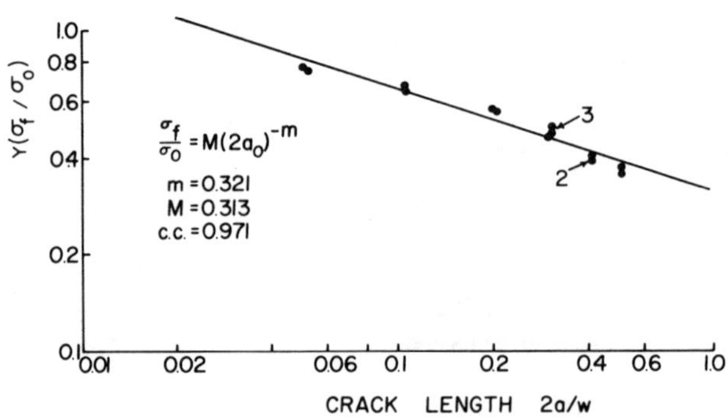

Figure 7. Experimental correlation for ML model.

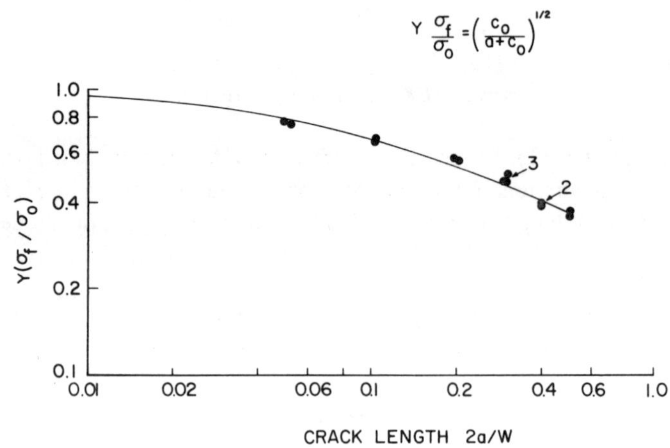

Figure 8. Experimental correlation for WEK model.

Crack-Tip Damage Growth and Failure Modes

As mentioned above, the rapid changes in COD are due to fiber failure at the crack tips which subsequently causes longitudinal plastic deformations and splitting in the matrix, shown in Figure 9a. This photomicrograph was taken after the specimen (crack length = 7.47 mm) was loaded up to 2.67 KN which is about 92% of its expected fracture strength. With increasing load, additional crack extension and additional longitudinal plastic zones are observed. The spacing between adjacent axial plastic zones equals approximately the fiber spacing. It should be noted that the length of the longitudinal plastic zones never extended beyond the specimens' gage length into the tabs. The final fracture surface is fairly irregular (Figure 9b, 10, 11b), resulting from a combination of the longitudinal plastic deformation of the matrix and the fracture of fibers at some weak spots along their lengths [11, 12]. This process repeats itself several times up to failure. The actual cracking of the matrix at the crack tip, Figure 10, did not take place during loading but only during the final stage of failure, probably because of the shock occurring during fracture. The splitting of the matrix runs through the entire thickness of the specimen, across the crack tip and around the fibers, Figure 9c. A view of the surface of the split matrix after fracture is shown in Figure 9d, revealing a large amount of shear plastic deformation.

It should be noted that the radius of the notch tip, resulting from the EDM of the notch, was relatively large (about 0.006 in.). However, it has been shown [3] that the effect of different radii at the notch tip (0.0023 and 0.005 in.) has little, if any, effect on the notched strength of unidirectinal B/Al, and even less effect when blunting of the crack tip is involved in the failure process, as is the case in our experiments.

The extension of the crack tip damage, as well as the number of longitudinal shear plastic zones, increases with load, as shown in Figure 12. These photographs were taken with an optical interferometry technique which enables measurement of the deformation zone immediately ahead of the crack tip [12]. In the present program, a collimated beam of monochromatic sodium vapor light ($\lambda = 0.5833 \mu m$) was directed through an optically flat quartz plate positioned directly on the polished specimen surface. During loading, the separation between the specimen's surface and the flat quartz plate, resulting from out-of-plate deformation at the crack tip, produces an optical interference pattern. No attempt has been made to obtain an optically flat surface, which is difficult to achieve because of the thinness of the specimens (0.052 in.) and the possibility of damaging the fibers immediately beneath the aluminum surface in polishing. Therefore, the quality of the pictures is not ideal [12], yet it is sufficient to identify the crack extension and the formation of new longitudinal plastic shear deformation bands of the matrix. As can be seen in Figure 12, the first longitudinal plastic deformation band is formed near the crack tip at relatively high load, about 80% of the notched strength. Crack extension then appears as the load increases further, and the whole process repeats itself.

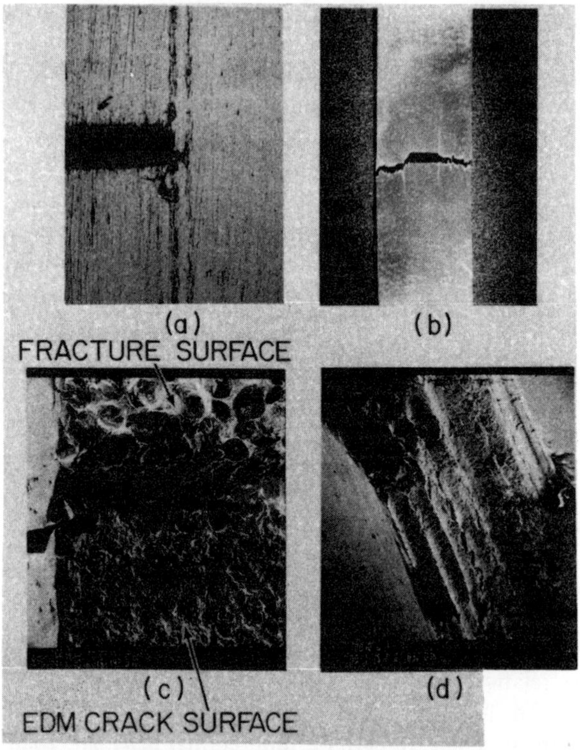

***Figure 9.** Crack tip damage growth and failure modes. a. Photomicrograph of crack tip damage showing crack extension and longitudinal plastic deformation. Crack length = 7.47mm. b. Photograph of fractured specimen showing matrix splitting at the crack tip. Crack length = 7.59mm. c. SEM view of matrix splitting at the crack tip. Crack length = 8.26mm. d. SEM side view of the split surface showing large amount of plastic deformation. Crack length = 7.72mm.*

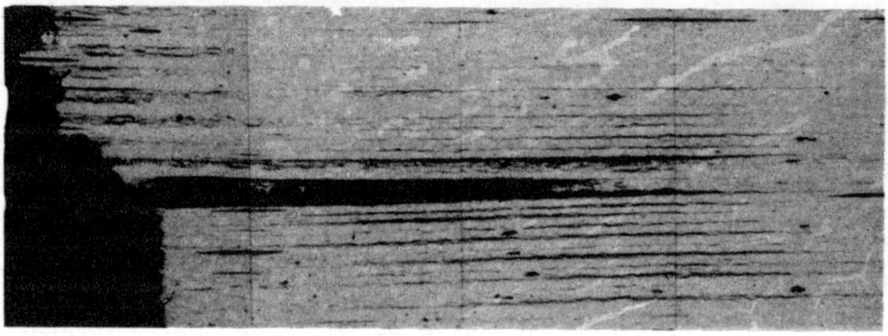

***Figure 10.** Photomicrograph of crack tip damage after fracture. Multiple longitudinal plastic deformation of the matrix material extends to the specimen's edge.*

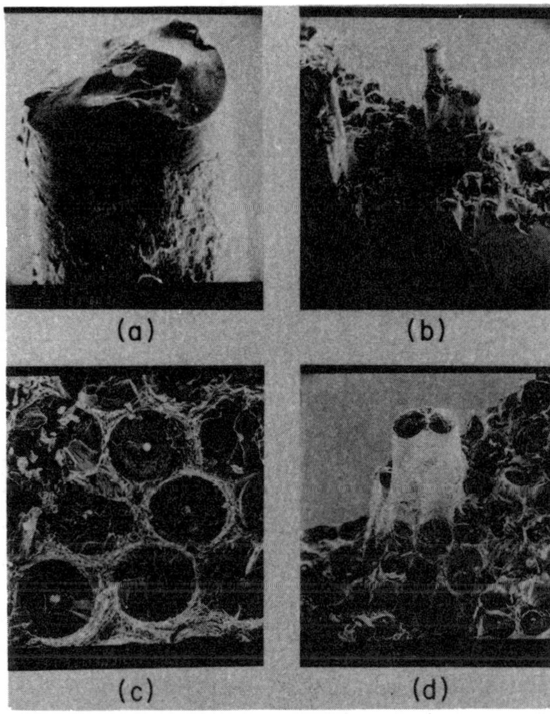

Figure 11. SEM photographs demonstrating good fiber-matrix and matrix-matrix bonding. a. – Individual filament; b, c, d – Fracture surface.

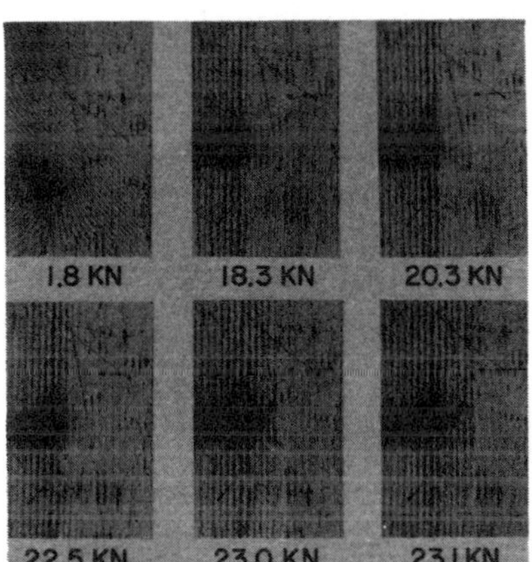

Figure 12. Optical interferometry photographs of B/Al (crack length = 8.26mm). under load. Crack extension and plastic zone increase with load.

Prior to fracture about 5 to 6 such shear yield zones were observed in several specimens of crack lengths varying from 0.3 to 0.5 inches. The plastic deformation of the matrix serves as a crack arresting medium, and additional crack extension alternates with the completion of successive longitudinal plastic deformation bands.

Many works studying the fracture behavior of unidirectional B/Al report that no slow crack growth actually occurs, but rather a sudden fracture is the typical failure mode. However, by properly polishing the test specimens and using the optical interferometry technique (or a standard interference microscope), stable crack tip damage growth can be detected as shown here. In fact, such an observation could be made through the load-COD curves as well. We also observed such stable damage growth for other metal matrix composite systems such as unidirectional borsic/titanium [23], B/Al-1100 [19], and B/Al laminate $[0/\pm22/0]_s$ [24]. Radiographs of preloaded specimens also confirmed this observation.

Since the spacing between adjacent plastic zones equals the fiber spacing, and since the matrix plastic deformation takes place prior to each crack extension (fiber failure), it is estimated that the total damage growth prior to failure (Δa) corresponding to 5 to 6 plastic zones at the crack tip amounts to about 0.7 to 0.8mm, in very good correlation with the estimated threshold crack extension given in [3]. Since the notched strength is mostly affected by the actual crack extension, it is of interest to compare this value (0.7 to 0.8mm) to the inherent flow size c_o in Equation (1) determined from the fracture strength results, Table 2. The comparison shows that c_o (= 1.0mm) is quite close to the actual extension of the crack Δa (= 0.7 to 0.8mm), while a_o in Equation (3) is quite different. Generally, the crack tip critical zone size c_o (or a_o), as introduced in [6, 7], is not based on the actual damage growth but is an empirical algebraic constant. However, it does serve very well its purpose of predicting the notched strength of unidirectional B/Al composites.

The damage growth, crack extension, and formation of longitudinal shear plastic zones depend also on the constituent properties, the interface properties, and the lay-up. For more ductile matrices, such as boron/aluminum-1100, the plastic deformation of the matrix was even observed at load levels as low as about 10% of the notched strength [19]. In such case, the crack arresting mechanism is more efficient and thus a lesser crack sensitivity is expected. Weak fibers combined with a strong, tough matrix seem to result in less notch sensitivity. For example, in borsic/titanium where the interphase effectively weakens the fibers considerably, notch sensitivity was found to be very small [23]. In B/Al laminate $[0/\pm22/0]_s$, crack blunting was negligible and a very high notch sensitivity was observed [24]. A detailed discussion on the metallurgical aspects of failure in B/Al composites is given, for example, in [25, 26].

The material tested in this program proved to be of high quality, having a very high longitudinal strength (2004 MPa). Scanning electron micrographs, Figure 12, indicate a very good fiber-matrix bond. Figure 12a shows a single fiber protruding from the matrix covered with a substantial residue of aluminum. Generally, the fibers are covered with residues of aluminum, Figure 12b, indicating a good fiber-matrix bond, i.e. the shear strength of the bond exceeds that of the aluminum matrix. Very little fiber pull-out has been observed. The few fibers shown in Figures 12b and 12d that have been pulled out of the opposite surface are exceptional and are shown here only to demonstrate the quality of the fiber-matrix bond. No matrix-matrix debonding has been observed, Figure 12c, and the large amount of microvoid formation (local yielding) seen appears to indicate the increased degree of triaxial restraint which occurs as the matrix-matrix bonding is enhanced [26]. The substantial shattering of the fibers, Figure 12c, as well as breaking of the tungsten core, is believed to be limited to a few fiber-diameters along the length of the fiber [23, 25, 26]. In some specimens the matrix material was dissolved in order to inspect the fibers in regions removed from the fracture surface. SEM inspections show that no multiple fiber failures have occurred in those regions.

Prediction of Load-COD Curves

The load-COD curve obtained by the IDG for various crack lengths, Figure 2, have been compared with an analytical model accounting for the longitudinal plastic deformations at the crack tip. As discussed earlier, the failure of notched unidirectional B/Al is preceded by crack tip blunting due to plastic shear deformation of the matrix in the filament direction. An analytical model predicting the length of the plastic zone, R_p, Figure 13, and the crack opening displacement at the crack tip (ΔV^p) has been suggested in [27, 28]. According to [28] the length of the plastic zone, R_p is

$$R_p = \frac{\pi^2}{16}\left[\frac{G_{LT}}{E_L}\right]^{1/2}\frac{aY^2P^2}{\tau_m{}^2W^2B^2} \tag{7}$$

and the corresponding permanent opening of the crack tip is:

$$\Delta V^p = \frac{8}{\pi}\left[\tau_m/(G_{LT}E_L)^{1/2}\right]R_p \tag{8}$$

where τ_m is the shear yield stress of the matrix and E_L and G_{LT} are the longitudinal and in-plane shear moduli, respectively. Note that ΔV^p predicts *crack tip* opening displacement (CTOD) but it can be compared to the COD at the *center* of the crack, measured experimentally, because it was found that the crack surfaces do not deform plastically during crack opening. Thus, combining Equations (7) and (8) we obtain

$$\Delta V^p = \frac{\pi}{4}\,\frac{1}{\tau_m E_L}\left(\frac{2a}{WB^2}\right)Y^2P^2 \tag{9}$$

The elastic COD at the center of the crack, ΔV^e, has been shown to be [13]:

$$\Delta V^e = \frac{2\alpha Y}{E_L B}\,\frac{2a}{W}\,P \tag{10}$$

where α is the orthotropic correction factor:

$$\alpha = \frac{1}{\sqrt{2}} \left[\left(\frac{E_L}{E_T} \right)^{1/2} - \nu_{LT} + \frac{E_L}{2G_{LT}} \right]^{1/2} \tag{11}$$

It has been shown [13] that the prediction of the elastic COD, Equation (10), agrees very well with the experimental results.

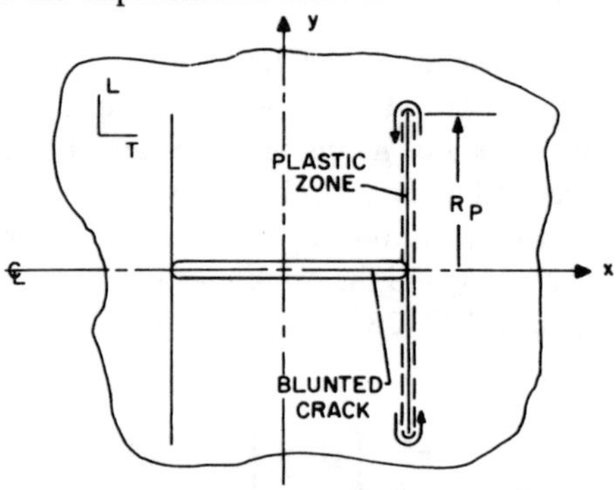

Figure 13. *Blunted crack and longitudinal plastic zone.*

The total crack opening displacement will be the sum of the elastic and plastic COD, i.e.:

$$\text{COD} = \Delta V^e + \Delta V^p = \frac{Y}{E_L B} \frac{2a}{W} \left[2\alpha + \frac{\pi Y}{4\tau_m W B^2} P \right] P \tag{12}$$

Equation (12) is compared very favorably with the experimental load-COD curves in Figure 14. The dashed lines in the Figures represent the initial slope of the load-COD curve as predicted from Equation (10). In order to achieve the good correlation between experiment and analytical model, the matrix shear yield stress τ_m was taken to be 207MPa (30ksi). In fact, the handbook data for boron/aluminum 6061-F give the value of 138MPa (20ksi) for τ_m. However, the tensile test data of the matrix material itself (fabricated of aluminum matrix alloy without the fiber reinforcement) show properties very close to aluminum 6061-T6 condition, which justify the choice of τ_m being equal to 207MPa (30ksi).

Coming back to Figure 14, we observe that the prediction deviates slightly from the experimental points at very low load levels. This is mainly due to the fact that the analytical model, Equation (12) predicts the formation of the longitudinal plastic shear deformation, and thus blunting of the crack tip, immediately upon initial loading. However, the first axial plastic zone actually appears at a much later

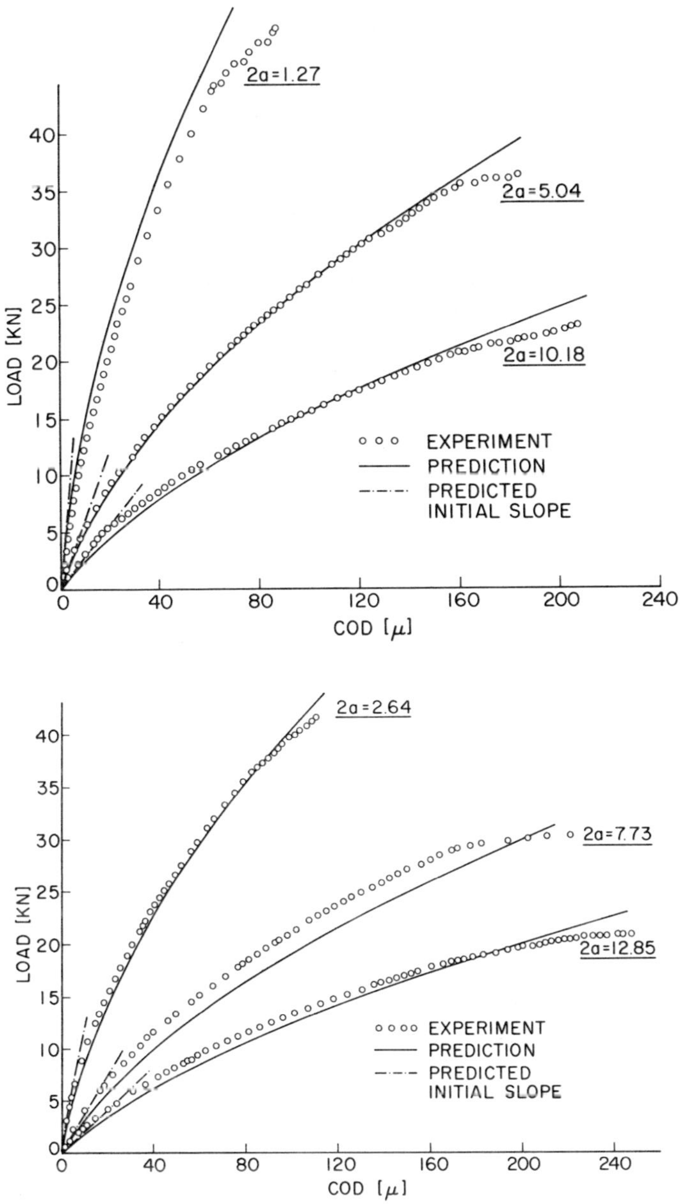

Figure 14. Comparison of experimental and analytical results of load-COD curves for various crack lengths

stage of the loading, as shown in Figure 12. The deviation of the prediction from the experimental data becomes more pronounced at higher load levels because the analytical model incorporates only one plastic zone, while at least 5 to 6 such zones are actually observed prior to failure, Figure 12. Therefore, Equation (12) predicts smaller crack openings than are actually observed as the load approaches its ultimate level.

As explained previously, after each intermediate loading, specimens were removed from the testing machine and the permanent opening of the crack was measured. The permanent opening of the crack was also determined from the loading-unloading curves, Figure 1. The experimental results of the permanent COD are compared with the prediction, Equation (9), as shown in Figure 15. The crack opening is given here in a nondimensional form in order to include results for the various crack sizes. The agreement between prediction and experiment is not as good, mainly because the prediction accounts for only one axial plastic zone rather than the multiple zones observed experimentally. However, the general trend of the data is very similar to the prediction.

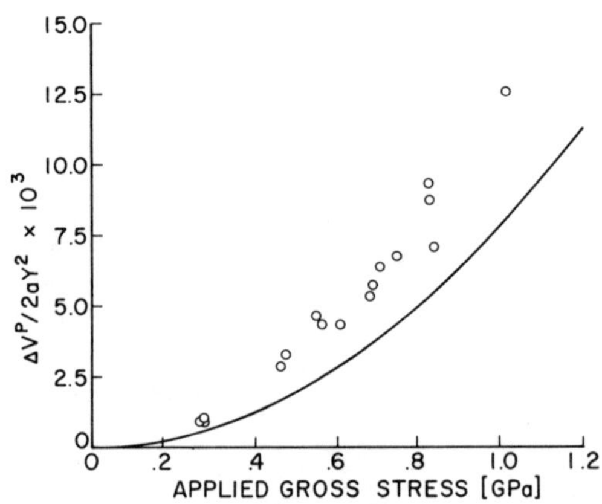

Figure 15. Comparison of experimental and analytical results of permanent COD as a function of applied load.

Resistance Curve Method

The resistance curve (K_R-Curve) method has been widely used to successfully characterize the fracture resistance of metals, e.g. [29, 30]. Application of this method to composites hinges on the assumption that the crack tip damage can be modeled as a self-similar crack extension and the effective crack length is determined through compliance matching. The question arises whether the resistance curve method can be applied to composite materials for which the damage growth can not be characterized as self-similar crack extension. Note that damage growth in composite laminates appears in the form of matrix cracking along the filament direction within individual plies, delamination, fiber breaks at weak locations, fiber-matrix debonding, etc. Obviously, the damage growth for unidirectional B/Al, as shown in this work, can hardly be characterized as self-similar crack extension. However, there have been some attempts to apply the resistance curve method to composites on the basis of compliance matching. Gaggar and Broutman [31] concluded that the K_R-curve concept can be adopted for epoxy and polyester composites randomly reinforced with discontinuous fibers. Morris and Hahn [17] applied the resistance curve method to $[0/\pm45]_s$, $[0/\pm45]_{2s}$, and $[0/90/\pm45]_s$ graphite/epoxy laminates, and they concluded that the K_R-curve can provide full information on the fracture resistance up to final failure. However, the results given in [17] indicate that the maximum value of K_R, as well as the amount of crack extension at failure, are not constant, and depend on the original crack length. Underwood [12] measured the out-of-plane deformation at the crack tip and the crack extension prior to failure using optical interferometry technique for single edge notched B/Al specimens. Using the K-calibration formula [20] for this geometry, the mode I stress intensity factor K_I was calculated and plotted against Δa. The data show a linear relationship between K_I and Δa for three different lay-ups: [0°], [0°/90], and [+60°/0°/−60°].

Another objective of this program is thus to investigate the applicability of the resistance curve approach to unidirectional B/Al composite. In particular, the effects of the test technique employed (IDG versus compliance gage) and of the loading history are studied.

The detailed procedure of plotting the resistance curve is well established and is given in details in [17, 29, 30] and will not be repeated in full here. Briefly, a straight line is drawn from the origin to a selected point of the load-COD curve, Figure 2, and the compliance is measured from the inverse of the secant slope. The value of this compliance is matched on the compliance curve (calibration curve) obtained for this material (see Figures 11 and 12 of Ref. [13]) and the effective crack length ($a_{eff} = a_o + \Delta a$) is determined. Repeating this procedure, the different values of the effective crack length as a function of the load are obtained. It should be emphasized here that the effective half crack length a_{eff} is not a pre-existing crack and Δa is not necessarily the actual self-similar crack extension, but it is some damage growth which can be modeled, through the compliance matching, as self-similar crack extension. The crack growth resistance K_R is obtained from:

$$K_R = Y\sigma_f\sqrt{\pi a_{eff}}\ ;\ a_{eff} = a_o + \Delta a \tag{13}$$

A typical K_R-curve obtained for center notch unidirectional B/Al is depicted in Figure 16. Three different K_R-curves are shown in the Figure:

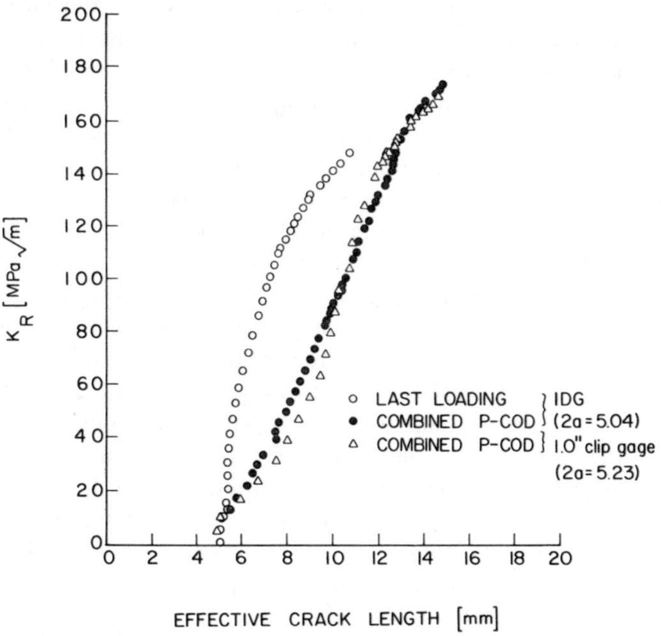

Figure 16. Fracture resistance data obtained from the IDG and compliance gage techniques.

(1) K_R-curve obtained from the "combined" load-COD curve, using the *local* compliance curve, Figure 12 of [13]---closed circles.

(2) K_R-curve obtained from "Last" Load-COD curve, using the *local* compliance curve, Figure 12 of [13]---open circles. Prior to last loading, the specimen was loaded to about 92% of its notched strength.

(3) K_R-curve obtained from load-displacement curve (compliance gage), using the *global* compliance curve, Figure 11 of [13]---open triangles.

All three curves are for 0.2 inch crack length. Comparing all three K_R-curves the following conclusions can be made:

1. The K_R-curve obtained by using the "combined" load-COD curve shows damage growth at a very early stage of the loading, due to the nonlinearity of the load-COD curve.

2. The K_R-curve obtained by using the "last" load-COD curve shows extension of damage at a much later stage of the loading. This is, however, misleading since some damage growth does exist prior to the "last" loading. Therefore, the maximum K_R is much smaller than the maximum value obtained from the "combined" load-COD curve.

3. The K_R-curve obtained by using the load-displacement curve (open triangles)

is very similar to the K_R-curve obtained by using the "combined" load-COD curve. This point, however, is not conclusive because such an agreement has not been observed for other crack lengths.

4. The point of instability of the critical fracture toughness K_C is defined by the point of tangency between K_R-curve as found from Equation (13), and the K_R-curve defined by

$$K_R = Y\sigma\sqrt{\pi a} \tag{14}$$

with σ as a parameter [29, 30]. Such a point of tangency could not be defined here; thus the last point on the K_R-curve defines K_C.

5. Comparison of the values of K_C from Figure 16 (K_C = 175MPa $\sqrt{m}$) and from Table 2 (K_C = 112.7MPa $\sqrt{m}$) shows a significant difference. The reason for this difference is mostly because the damage growth Δa used in Equation (13) is much longer than c_o as determined from the best-fit of the notched strength data, Figure 4. In fact, we could anticipate such a difference directly from the loading-unloading curves and the mode of crack tip damage growth. The longitudinal plastic deformation at the crack tip causes significant increase in COD, resulting in a large plateau in the load-COD curve.

6. Because of such a flat load-COD curve, using the compliance matching method for notches longer than 0.2 in. resulted in effective crack lengths which are longer than the width of the specimen! This is, of course, impossible. In other words, the K_R-curve cannot be applied to materials exhibiting the damage growth characteristic of unidirectional B/Al, and the assumption of "self similar crack extension" can not be applied when a sensitive displacement gage (such as the IDG) is employed.

7. It should be remembered that the change of the secant slope in the load-COD curve does not represent the actual crack extension. Since there remains permanent crack tip deformation upon unloading, Figure 1, the actual crack extension can be better characterized in terms of the change in the elastic compliance obtained from incremental loading-unloading curves. When this is done, a more reasonable K_R-curve will result for all crack lengths.

Figure 17 shows the K_R-curve obtained for the various initial crack lengths from the load-displacement curves (compliance gage). As this method is not as sensitive to damage growth as the IDG is, K_R-curves could be obtained for all crack lengths. The conclusions from Figure 17 are as follows:

1. The maximum K values measured from the K_R-curves are much higher than predicted using the WEK model, Equation (1).

2. The damage growth Δa is much longer than c_o and is not constant, as was also found in [17] for graphite/epoxy laminates. The maximum K values are not constant and depend on the initial crack length.

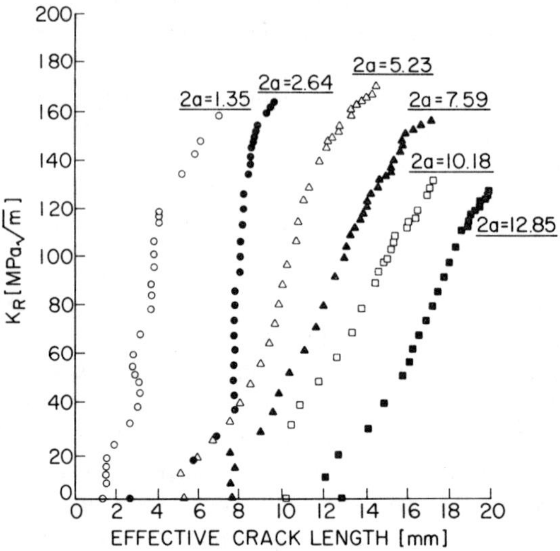

***Figure 17.** Fracture resistance data for various crack lengths obtained with compliance gage.*

CONCLUSIONS

Load-crack opening displacement (COD) curves were obtained for crack length-to-width ratios varying from 0.05 to 0.5. The COD was measured across the crack surfaces by means of the laser interferometric technique. This technique was found to be highly sensitive in detecting crack tip damage, very accurate on the submicron scale, and simple to utilize.

The load-COD curves for center notched unidirectional boron/aluminum are highly nonlinear. Upon unloading, a relatively large permanent crack opening displacement remains, mainly because of the plastic deformation of the matrix at the crack tip. Such a correlation between the observed damage growth and deformation characteristics could not be confirmed using the standard compliance or clip gages.

The fracture strength of the material can be predicted by using the existing analytical models based on the concept of a critical damage zone size c_o at the crack tip being a material parameter. It has been determined that this damage zone size, obtained from the best fit of the data, agrees reasonably well with the observed damage growth, but it is much different from the damage size determined by using the resistance curve method. However, the notch strength data obtained for B/Al are in good agreement with predictions. Another proposed model, in which $r^{-1/2}$ singularity is not assumed, does not follow the experimental data as well.

The load-COD curves obtained by the IDG method have been compared with an analytical model. The model incorporates the longitudinal plastic deformation which eventually leads to cracking of the matrix. An excellent agreement is established between predictions and experiments for all the crack sizes tested.

A detailed study of the damage growth and failure mode indicates that the crack tip damage appears in the form of plastic deformation of the ductile matrix in the longitudinal direction. Prior to failure, multiple longitudinal plastic zones appear, accompanied by crack extension. The severity of the crack tip damage is demonstrated by the high nonlinearity of the load-COD curve, by the large amount of COD, and by the large permanent displacement remaining upon unloading. The resistance curve method has also been tried. However, it was found that the resistance curves depend strongly on the test method (IDG versus compliance gage) as well as the loading history. The calculated stress intensity factor and effective increment of crack length, both at fracture, were found to depend on the initial crack length.

It appears that the fracture behavior of the subject composite material under uniaxial static tension can be predicted. Also, the test procedure employed in this study is very helpful in better understanding the fracture behavior of composite materials. However, it seems that a full characterization of the fracture behavior of composite materials depends strongly on the test procedure and the mode of failure, and that a specific analytical approach has to be employed for each material system.

ACKNOWLEDGMENTS

The work of the first author (J. Awerbuch), performed at the Air Force Materials Laboratory (AFML), Wright Patterson AFB, Ohio, was supported in part by the National Research Council through the Resident Research Associateship Program and in part by the Metal and Ceramic Division of AFML through contract No. F33615-76-C-5124 with the University of Dayton Research Institute.

Special thanks are due T. Nicholas of AFML for implementing this program and for high technical assistance and encouragement. The support of S.W. Tsai in our research program is also acknowledged with appreciation.

The authors wish to express sincere gratitude to Messrs. R. Klinger and C. Bell for their help in the experimental program. Special thanks go to Mr. D.E. Macha for his guidance in the use of the laser interferometric test method. The help of J.W. Jones, of the University of Michigan in applying the optical interferometry technique is deeply appreciated.

REFERENCES

1. E. L. Foster, Jr., "Technology Development of Metal Matrix Composites to DOD Application Requirements," proceeding of First MMC Workshop, Institute of Defense Analysis, paper P-1144, September 1975.
2. J. Awerbuch and H. T. Hahn, "Hard Object Impact Damage of Metal Matrix Composites," *J. Composite Materials*, Vol. 10 (1976), p. 231.
3. K. G. Kreider and L. Dardi, "Fracture Toughness of Composites," *Failure Modes in*

Composites, I. J. Toth Edt., The Metallurgical Society of the AIME, Vol. I, New York, New York, 1972, p. 193.

4. J. W. Mar and K. Y. Lin, "Fracture Mechanics Correlation for Tensile Failure of Filamentary Composites With Holes," *J. of Aircraft,* Vol. 14 (1977), p. 703.
5. M. A. Wright and F. A. Iannuzzi, "The Failure of Notched Specimens of Boron-Fiber Reinforced 6061 Aluminum Alloy," *Failure Modes in Composites,* J. N. Fleck and R. L. Mehan, Eds., The Metallurgical Society of the AIME, Vol. II, New York, New York, 1974, p. 68.
6. M. E. Waddoups, J. R. Eisenmann and B. E. Kaminski, "Macroscopic Fracture Mechanics of Advanced Composite Materials," *J. Composite Materials,* Vol. 5 (1971), p. 446.
7. R. J. Nuismer and J. M. Whitney, "Uniaxial Failure of Composite Laminates Containing Stress Concentrations" in Fracture Mechanics of Composites, ASTM STP 593, American Society of Testing and Materials, 1975, p. 117.
8. J. R. Hancock and G. D. Swanson, "Toughness of Filamentary Boron/Aluminum Composites," *Composite Materials: Testing and Design* (Second Conference), ASTM STP 497, American Society for Testing and Materials, 1972, p. 299.
9. C. T. Sun and K. M. Prewo, "The Fracture Toughness of Boron Aluminum Composites," *J. Composite Materials,* Vol. 11 (1977), p. 164.
10. W. R. Hoover and R. E. Allred,"The Toughness of Borsic-Aluminum Composites With Weak Fiber Matrix Bonds," *Failure Modes in Composites,* I. J. Toth Edt., The Metallurgical Society of the AIME, Vol. I, New York, New York, 1972, p. 311.
11. K. M. Prewo and K. G. Kreider, "Fatigue Failure Mechanism in Boron Aluminum," *Failure Modes in Composites,* I. J. Toth Edt., The Metallurgical Society of the AIME, Vol. I, New York, New York, 1972, p. 395.
12. J. H. Underwood, "Crack-Tip Deformation Measurements Accompanying Fracture in Fibrous and Laminar Composites," *Composite Materials: Testing and Design* (Third Conference), ASTM STP 546, American Society for Testing and Materials, 1974, p. 376.
13. J. Awerbuch and H. T. Hahn, "K-Calibration of Unidirectional Metal Matrix Composites," *J. Composite Materials,* Vol. 12 (1978), p. 222.
14. D. E. Macha, W. N. Sharpe Jr., and A. F. Grant Jr., "A Laser Interferometry Method for Experimental Stress Intensity Factor Calibration," *Cracks and Fracture,* ASTM STP 601, American Society for Testing and Materials, 1976, p. 490.
15. W. N. Sharpe Jr., "Interferometric Surface Strain Measurements," *International J. Nondestructive Testing,* Vol. 3 (1971), p. 56.
16. J. P. Waszcak, "Applicability of Linear Elastic Fracture Mechanics to 5.6 mil. Boron/6061 Aluminum," *J. Aircraft,* Vol. 13 (1976), p. 770.
17. D. H. Morris and H. T. Hahn, "Fracture Resistance Characterization of Graphite/Epoxy Composites," *Composite Materials: Testing and Design* (Fourth Conference), ASTM STP 617, American Society for Testing and Materials, 1977, p. 5.
18. I. M. Daniel, R. E. Rowlands and J. B. Whitside, "Deformation and Failure of Boron/Epoxy Plate with Circular Holes," *The Test Methods for High Modulus Fibers and Composites,* ASTM STP 521, American Society for Testing and Materials, 1973, p. 143.
19. J. Awerbuch, "Failure Mechanism and Deformation Characteristics of Unidirectional Boron Aluminum Composites," to be published.
20. W. F. Brown and J. E. Srawley, in *Plane Strain Crack Toughness Testing of High Strength Metallic Materials,* ASTM STP 410, American Society for Testing and Materials, 1966.
21. R. L. Crane, Nondestructive Evaluation Branch at the Air Force Materials Laboratory, Wright Patterson AFB, Ohio, 45433. Private Communication.
22. E. M. Wu, in *Composite Materials Workshop,* S. W. Tsai, J. C. Halpin and N. J. Pagano, Eds., Technomic Publishing Co., Inc., 1968.
23. J. Awerbuch and H. T. Hahn, "Crack-Tip Damage and Fracture Toughness of Borsic/Titanium Composite," to be published.

24. J. Awerbuch, "Notch Sensitivity of Boron Aluminum Laminate," to be published.
25. R. C. Jones, "Fractography of Aluminum-Boron Composites," *Composite Materials: Testing and Design* (First Conference), ASTM STP 460, American Society for Testing and Materials, 1969, p. 512.
26. R. C. Jones and J. L. Christian, "Analysis of an Improved Boron/Aluminum Composite," *Composite Materials: Testing and Design* (Second Conference), ASTM STP 497, American Society for Testing and Materials, 1972, p. 439.
27. J. Tirosh, "The Effect of Plasticity and Crack Blunting on the Stress Distribution in Orthotropic Composite Materials," *J. Appl. Mech.*, Vol. 40 (1973), p. 785.
28. F. A. McClintock, "Problems in Fracture of Composites with Plastic Matrices," presented at the Fourth Symposium on High Performance Composites at St. Louis, MO., Monsanto Co. and Washington University, 1969.
29. *Fracture Toughness Evaluation by R-curve Methods*, ASTM STP 527, American Society for Testing and Materials, 1973.
30. J. Eftis, D. L. Jones and H. Liebowitz, Fracture Mechanics of Aircraft Structures, H. Liebowitz, Ed., AGARD-AG-176, North Atlantic Treaty Organization, 1974, p. 32.
31. S. Gaggar and L. J. Broutman, "Crack Growth Resistance of Random Fiber Composites," *J. Composite Materials*, Vol. 9 (1975), p. 216.

Notched Strength of Composite Materials*

R. BYRON PIPES

University of Delaware
Center for Composite Materials
Newark, DE 19711

ROBERT C. WETHERHOLD

E. I. duPont de Nemours Company (Inc.)
Engineering Department
Wilmington, DE 19898

AND

JOHN W. GILLESPIE, JR.

University of Delaware
Center for Composite Materials
Newark, DE 19711

(Received July 5, 1978)

ABSTRACT

A macroscopic model for predicting the strength of a composite laminate containing a circular notch is introduced. A property, which quantifies the reduction in strength of a given composite material or laminate due to a circular notch, is proposed. The superposition of notched strength data for several important composite material systems and laminate stacking sequences is achieved through development of a notched strength-radius superposition method and the definition of radius shift parameters, a_c and a_m. The superposition allows prediction of the influence of laminate stacking sequence upon notched strength. In addition,

---

*Partially supported by the Naval Air Development Center under contract N62269-76-R-0167

data for composite material systems or laminates of common stress concentration factor is shown to superimpose by use of the developed shift parameters.

SCOPE

COMPOSITE LAMINATES CONTAINING cutouts or penetrations offer a special challenge to the designer since design by the classical stress concentration factor is unnecessarily conservative. This is due to the fact that for high modulus composite materials, laminate strength has been found to be a strong function of the absolute dimension of the cutout. In addition, this "hole size effect" is more pronounced for materials of significant orthotropy and composed of relatively brittle matrices.

To date two methods for prediction of the tensile strength of notched (circular hole or slit) composite laminates have evolved. The first method employs the principles of linear elastic fracture mechanics (LEFM), while the second method focuses upon the finite state of stress near the boundary of the notch or cutout. Waddoups et al [1] and Tirosh [2] have proposed fracture mechanics models which contain a hypothetical intense energy region of dimension "a" containing a Griffith crack adjacent to the machine notch. The characteristic dimension is assumed to be a material property and to remain constant for all values of notch radii. Thus, the model proposed is identical to the LEFM model of the circular hole containing two edge cracks of length "a." Waddoups employs the development by Bowie [3] for an isotropic material to determine the function $f\,(a/R)$ where "R" is the notch radius. In addition, Waddoups employs an isotropic finite-width correction factor to account for the effects of finite boundaries upon test results. The LEFM models contain two undetermined parameters: the characteristic dimension "a" and the unnotched tensile strength of the laminate, "σ_o." Having determined "σ_o" and the value of "a" for one notch size, the model allows prediction of the strength of laminates for any other notch size.

However, several limitations of the Waddoups model preclude its general utility. First the necessary development by Bowie [3] is limited to isotropic material systems. In addition, the assumption that the characteristic dimension is a material constant is invalid as will be shown later.

The second method for predicting notched laminate tensile strength was developed by Whitney and Nuismer [4–5] as an alternative to the LEFM model. This model requires the knowledge of the finite stress distribution in the vicinity of the notch. Failure is assumed to occur when the stress at a distance "d_o" from the edge of the notch reaches the unnotched tensile strength of the laminate, "σ_o." Since Whitney and Nuismer took "d_o" to be constant for all notch sizes, their formulation yielded a two parameter model which was essentially equivalent to the LEFM models in prediction of the dependence of notched strength upon notch radius.

The objective of the present paper is to introduce a new three-parameter strength model, to discuss the physical significance of the model parameters, and to demonstrate the utility of this new model in predicting the notched strength of unidirectional boron-aluminum and various graphite-epoxy laminates.

NOTCHED STRENGTH MODEL

The influence of the radius, R of an unloaded circular hole upon the strength of a laminate of infinite extent is given in the following relation: [1, 4, 6].

$$\sigma_N/\sigma_o = 2\ \{2 + f(R)^{-2} + 3f(R)^{-4} - (K_T^\infty - 3)\ [5f(R)^{-6} - 7f(R)^{-8}]\}^{-1} \tag{1}$$

where

$$f(R) = [1 + R^{m-1} R_o^{-m} C^{-1}] \tag{2}$$

σ_o = unnotched laminate strength
σ_N = notched laminate strength

The stress concentration factor K_T^∞ for a given laminate or material can be expressed as a function of the effective orthotropic material properties.

$$K_T^\infty = 1 + \sqrt{2(\sqrt{E_y/E_x} - \nu_{yx}) + E_y/G_{yx}} \tag{3}$$

where E_x, E_y, ν_{yx}, G_{yx} are effective orthotropic engineering elastic constants of the laminate or material, and y is the loading axis. The undetermined parameters in the model are C, the "notch sensitivity factor," and m, the "exponential parameter."

It is now informative to examine separately the influence of three variables in equation (1) which influence notched strength: notch sensitivity factor, C; exponential parameter, m; and stress concentration factor, K_T^∞. The influence of notch sensitivity factor upon notch strength is shown in Figure 1 where increases in notched sensitivity factor result in reduction in strength for a given notch radius. Indeed notch sensitivity factor shifts these curves along the radius axis without distorting the shape of the curve. Note that for complete notch insensitivity ($C = 0$), notched strength is equal to unnotched strength for all notch radii and for infinite notch sensitivity ($C = \infty$), notched strength/unnotched strength is equal to $1/K_T^\infty$.

The exponential parameter "m" acts to change the slope of the notch sensitivity curve as shown in Figure 2. This parameter should be $0 \leqslant m < 1$ for admissible material systems. The upper bound value $m = 1$ results in a realtion (1) which is independent of notch radius, while for the lower bound $m = 0$, relation (1) reduces to the two parameter model [1, 4].

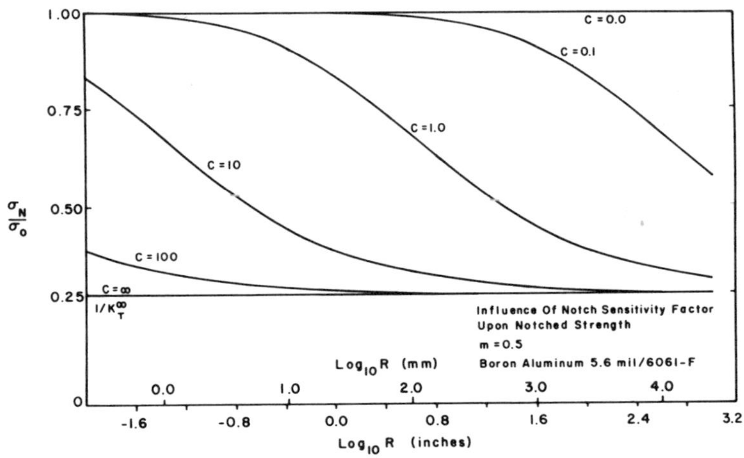

Figure 1. Influence of notch sensitivity factor upon notched strength.

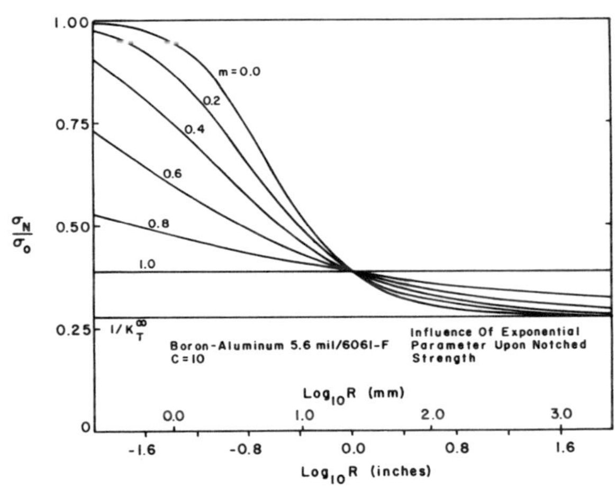

Figure 2. Influence of exponential parameter upon notched strength.

The influence of the stress concentration factor K_T^∞ is important, in general, for large notch radii. As the notch radius approaches infinity, notched strength converges to σ_o/K_T^∞. However, most important composite materials and laminates which exhibit mode I fracture, possess a stress concentration factor between 2.0 and 4.0 as shown in Table 1. Figure 3 shows that for notch radii of less than 1.00 inches, the notch sensitivity relations very nearly coincide for intermediate values of notch sensitivity factor ($C \approx 10$ in^{-1}).

Table 1. Stress Concentration Factor for Various Composite Materials

Material	Laminate	K_T^∞
intermediate-modulus	$[\pm 45]_s$	2.060
graphite-epoxy	$[0/\pm 45/90]_s$	3.000
(AS/3501)	$[0_2/\pm 45]_s$	3.483
	$[0_4/\pm 45]_s$	4.074
high modulus	$[\pm 45]_s$	1.806
graphite-epoxy	$[0/\pm 45/90]_s$	3.000
(GY 70)	$[0_2/\pm 45]_s$	3.540
boron-epoxy	$[\pm 45]_s$	1.851
(AVCO 5505)	$[0/\pm 45/90]_s$	3.000
	$[0_2/\pm 45]_s$	3.494
boron-aluminum	$[\pm 45]_s$	2.706
(5.6 mil/6061-F)	$[0/90/\pm 45]_s$	3.000
	$[0_2/\pm 45]_s$	3.122
	$[0_6]$	3.607*
glass-epoxy	$[\pm 45]_s$	2.382
(3M Scotchply)	$[0/90/\pm 45]_s$	3.000
	$[0_2/\pm 45]_s$	3.313

*Material investigated in current study

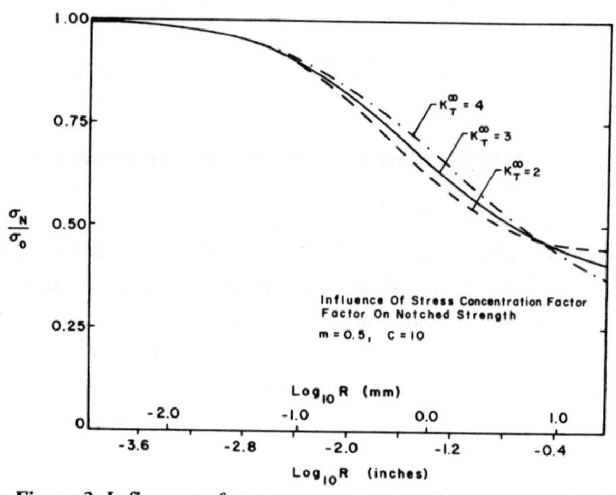

***Figure 3.** Influence of stress concentration factor on notched strength.*

PARAMETERS INFLUENCING NOTCHED STRENGTH

Several parameters influence notched strength in composite materials. Among these are fiber-matrix interfacial bonding, matrix properties, fiber properties, fiber orientation and laminate stacking sequence. In the present model all of these influences are represented by the two material properties: notch sensitivity factor "C" and the exponential parameter "m." In the present work we focus upon two of the above parameters to demonstrate their influence and to show the facility of the model for representation of the effect of these parameters upon notched strength.

First, consider the influence of insitu matrix properties upon the notched strength of the boron-aluminum composite, 5.6 mil/6061-F. Unidirectional laminates consisting of 6 plys were fabricated under two different processing conditions. The insitu matrix properties were determined from aluminum foil laminates subjected to identical processing conditions. The two composites produced were of identical volume fraction and contained identical fiber types. However, due to differences in the thermal processing cycles, the matrix phases in the two composites were not identical, but differed in yield strength. The processing conditions for the two composites are given as follows:

Material I

Heat to 752°F (400°C) for one hour to exhaust resin binder. Diffusion bonding is then carried out at 975°F (496°C) and 5000 psi (35 MPa) for one-half hour followed by slow cooling. The measured yield stress of material I was 6.2 ksi (42.8 MPa). The total process time excluding cooling was one and one-half hours.

Material II and III

The panel was heated to 500°F (260°C) for one-half hour and then to 785°F (418°C) for two and one-half hours to exhaust the resin binder. Diffusion bonding was accomplished at 975°F (496°C) and 5000 psi (34.5 MPa) for one and one-half hours. The pressure was reduced to atmospheric and the panel was cooled in the press for one hour. The panel was removed from the press for continued cooling at ambient conditions. The total processing time, excluding cooling, was four and one-half hours. The measured yield stress of material II was 4.2 ksi (29 MPa).

It is significant to note that while the matrix yield strengths of the two composites were different, their unnotched tensile strengths were identical (225 ksi, 1.55 GPa). Hence, it would appear that optimization of process conditions to yield maximum unnotched tensile strength without regard to notch sensitivity would not lead to an optimized structural material. An important parallel may be drawn here to past efforts to maximize yield strength in homogeneous metals while ignoring the deleterious influence of these efforts upon fracture toughness.

The influence of processing conditions upon composite notched strength is shown graphically in Figure 4 and summarized in Table 2. These results indicate

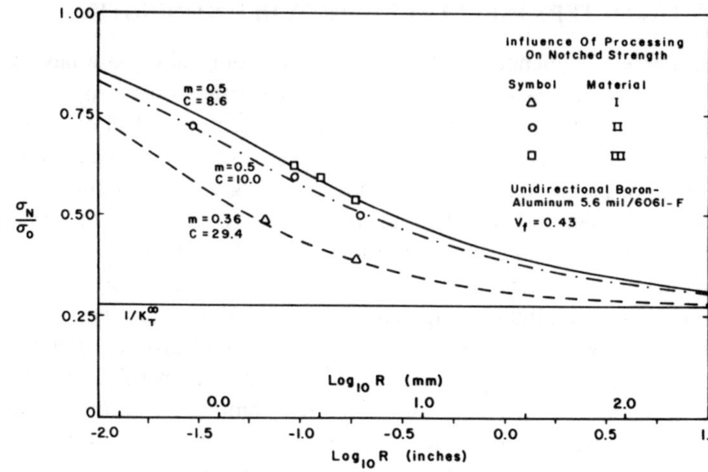

Figure 4. Influence of processing on notched strength.

Table 2. Notched Strength Data for Boron-Aluminum 5.6 Mil/6061-F

Material	R (in/mm)	$\log_{10}R$ (in/mm)	Average σ_N/σ_o	Average $(K_T/K_T^{\infty})^*$ σ_N/σ_o	Average σ_o (ksi/MPa)
I	0.068/1.73	-1.17/0.238	0.483	0.485	225/155
I	0.189/4.80	-0.72/0.681	0.364	0.388	"
II	0.030/0.76	-1.52/-0.12	0.719	0.719	"
II	0.096/2.44	-1.02/0.39	0.586	0.595	"
II	0.198/5.03	-0.70/0.70	0.466	0.500	"
III	0.095/2.41	-1.02/0.38	0.613	0.622	"
III	0.127/3.23	-0.90/0.51	0.576	0.592	"
III	0.188/4.78	-0.73/0.68	0.503	0.537	"

*Finite-Width Correction; K_T calculated by finite-element techniques

that material I exhibited a much greater notch sensitivity ($C_I = 29.4$ in^{-1}) than materials II and III ($C_{II} = 10$ in^{-1}, $C_{III} = 8.6$ in^{-1}). Although the process conditions for materials II and III were recorded as identical, these results indicate that the two materials, processed at different times, were not processed in exactly the same manner. The exponential parameter "*m*" was determined to be 0.5 for materials II and III and 0.36 for material I as shown in Figure 4.

The second parameter considered in the present study was laminate stacking sequence. Earlier investigations [8] have shown that laminate strength is a function of stacking sequence due to interlaminar edge effects. More recent studies of

Whitney and Kim [9] have shown that notched strength is also a strong function of laminate stacking sequence. If the data summarized in Table 3 are examined in light of the definition of notch sensitivity factor, it is found that for the $[\pm 45/0/90]_s$ laminate, $C = 7.5 \text{ in}^{-1}$ and $m = 0.475$, while for the $[90/0/\pm 45]_s$ laminate, $C = 11.5 \text{ in}^{-1}$ and $m = 0.40$. Indeed the $[90/\pm 45/0]_s$ laminate exhibits the greater notch sensitivity as shown in Figure 6. In addition it is significant to note that the parameter "m" was not found to be constant for the two stacking sequences. While these data are admittedly limited, they do seem to refute the existence of a universal value of "m" as suggested by Karlak [6].

Table 3. Whitney-Kim Notched Strength Data for Quasi-Isotropic Laminates [9] T300/934

Laminate	R (in/mm)	$\log_{10}R$ (in/mm)	$(K_T/K_T^\infty)^* \sigma_N/\sigma_o$	σ_o (ksi/MPa)
$[\pm 45/0/90]_s$	0.05/1.27	-1.30/0.104	0.721	67.61/446
"	0.15/3.81	-0.82/0.581	0.595	"
"	0.30/7.62	-0.52/0.882	0.514	"
$[90/0/\pm 45]_s$	0.05/1.27	-1.30/0.104	0.667	79.94/551
"	0.15/3.81	-0.82/0.581	0.538	"
"	0.30/7.62	-0.52/0.882	0.464	"

*Finite-Width Correction based on isotropic correction factor

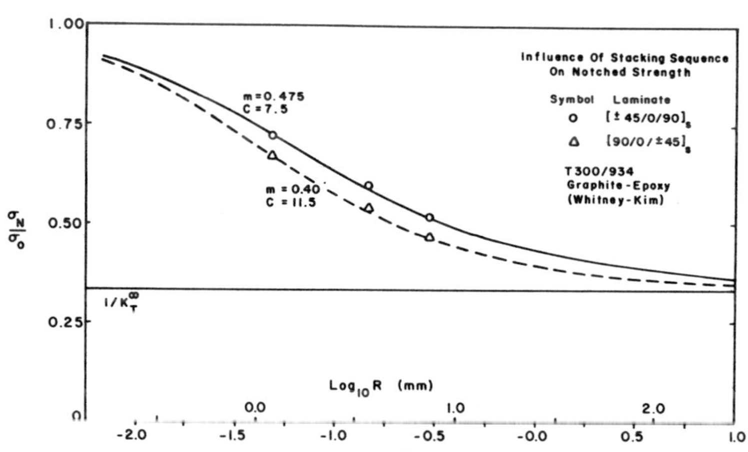

***Figure 5.** Influence of stacking sequence on notched strength.*

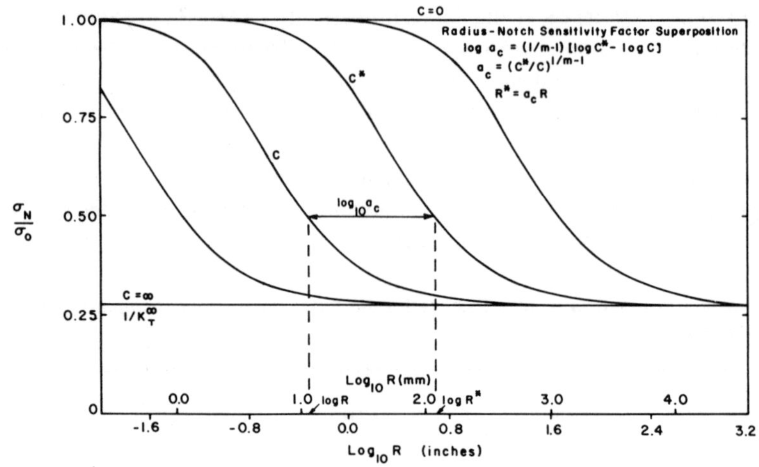

Figure 6. Radius-notch sensitivity factor superposition.

NOTCHED STRENGTH – RADIUS SUPERPOSITION METHOD

When the stress concentration factor K_T^∞ is constant for a given composite material system or system of laminate configurations, it is possible to superimpose all the notched strength data representing various values of notch sensitivity "C" and exponential parameter "m" to a single master curve. This is accomplished by drawing an analogy to the time-temperature superposition principle of linear visco-elasticity [10, 11]. Letting the reference radius $R_o = 1''$, a radius shift parameter "a_c" is defined as follows:

$$a_c = (C^*/C)^{\frac{1}{m-1}} \quad (0 \leqslant m < 1) \tag{4}$$

The notched strength data for a notch sensitivity, C is then superimposed upon that of notch sensitivity, C^* by shifting the strength at radius, R to R^* as shown in Figure 6

$$R^* = a_c R \tag{5}$$

When the strength data for materials or laminates of different exponential parameters are to be superimposed a second radius shift parameter is defined:

$$a_m = \frac{m-1}{m^*-1} \quad (0 \leqslant m, m^* < 1) \tag{6}$$

The notched strength data for a given exponential parameter, m is superimposed upon that for m^* by shifting the strength at radius R to R^* as illustrated in Figure 7.

$$\log_{10} R^* = a_m \log_{10} R \tag{7}$$

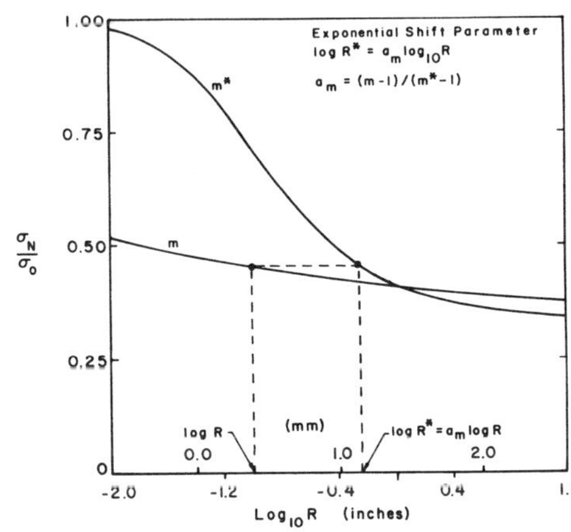

Figure 7. Exponential shift parameter.

In this way, it is possible to construct notched strength as a function of notch radius for all such laminate configurations or materials by measuring the relation for a single laminate or material and measuring sufficient data (typically three points) for the new material or laminate in order to establish the shift parameters.

It should be noted that the superposition principle allows quantitative characterization of the influence of laminate stacking sequence upon notched strength since K_T^∞ does not vary with laminate sequence. In addition, it is possible to construct relations for notched strength versus hole size for any stacking sequence given that for at least one other stacking sequence and the strength for two hole sizes in the desired stacking sequence. Figure 8 shows the data generated by Whitney [9] and Waddoups et al [1] shifted by the superposition method to form the master curve for the $[\pm 45/0/90]_s$, $[90/0/\pm 45/]$ and $[0/\pm 45/]_{2s}$ laminates at $C = 10\ \text{in}^{-1}$, $m = 0.5$, $K_T^\infty = 3.0$. Similarly, the strength-hole radius relations for the boron-aluminum material subjected to varying process conditions can be combined to form a single master curve for that material system. These results are presented in Figure 9 where the data are superimposed to a master curve $C = 10\ \text{in}^{-1}$, $m = 0.5$, $K_T^\infty = 3.607$.

In examinig radius-notch sensitivity superposition, it is instructive to note the influence of the stress concentration factor K_T^∞. As stated earlier, K_T^∞ must remain constant for the materials or laminates considered in order for the superposition

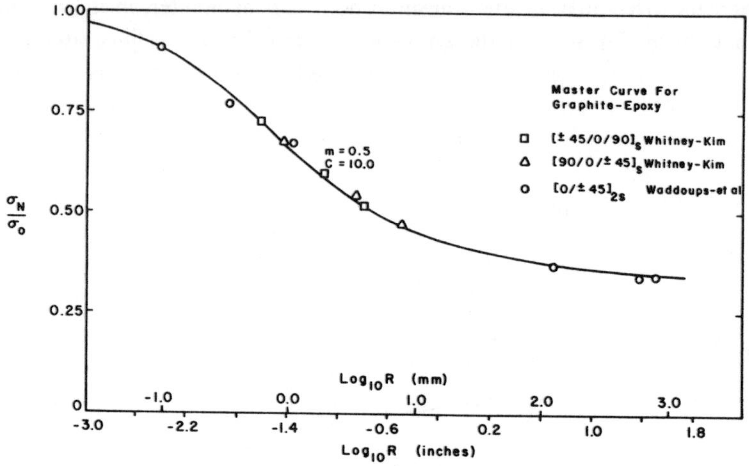

Figure 8. Master curve for graphite-epoxy.

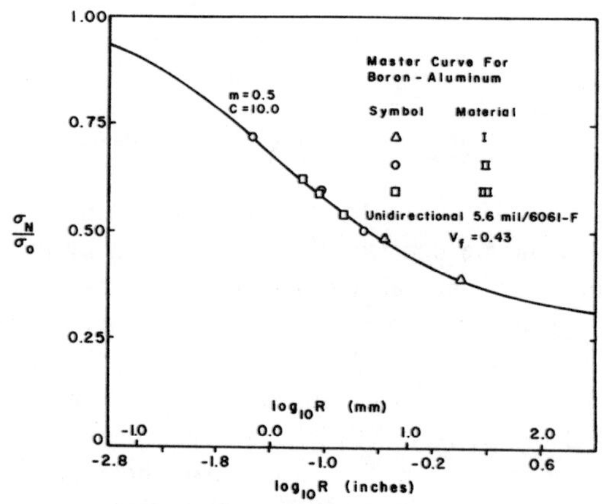

Figure 9. Master curve for boron-aluminum.

principle to be valid. It would appear that this restriction limits the class of materials or laminates in which the principle may be applied. This is in fact the case; however, the very important class of laminates termed quasi-isotropic possess a common stress concentration factor, $K_T^\infty = 3.0$. Hence, the method has general applicability to the quasi-isotropic laminates for all material systems.

CONCLUSIONS AND SIGNIFICANCE

The new strength model, as described in the present work, shows excellent agreement with experiment data for a wide range of composite material systems and laminates. The influence of the important model parameters have been discussed and their ability for representation of the influence of process conditions and laminate stacking sequence have been assessed. A parameter has been defined as "notch sensitivity factor," which appears to quantify the reduction in strength of a given composite material or laminate of infinite extent due to the presence of a circular notch.

A notched strength-radius superposition method has been developed which allows superposition of notched strength to a single master curve for all materials and laminates of common stress concentration factor, K_T^∞. The superposition method allows determination of the notched strength as a function of notch radius for all laminate stacking sequences, given the relation for a single stacking sequence by determining shift parameters, a_c or a_m notched strength data for quasi-isotropic laminates of all composite materials may be superimposed to form a single master relation since $K_T^\infty = 3.0$ for all such laminates.

NOMENCLATURE

Symbol	Definition
a	characteristic dimension
a_c	radius shift parameter for notch sensitivity factor
a_m	radius shift factor for exponential parameter
C,C^*	notch sensitivity factor
E_y	effective longitudinal Young's modulus
E_x	effective transverse Young's modulus
G_{yx}	effective inplane shear modulus
K_T^∞	infinite plate stress concentration factor
m	exponential parameter
R	circular notch radius
R_o	reference circular notch radius
ν_{yx}	effective Poisson's ratio
σ_o	unnotched laminate or material strength
σ_N	notched laminate or material strength

REFERENCES

1. M. E. Waddoups et al., "Macroscopic Fracture Mechanics of Advanced Composite Materials," *J. Composite Materials,* Vol. *5* (1971), p. 446.
2. J. Tirosh, "On the Tensile and Compressive Strength of Solids Weakened (Strengthened) by an Inhomogeneity," *J. Applied Mechanics,* September 1977, p. 449.
3. O. L. Bowie, "Analysis of an Infinite Plate Containing Radial Cracks Originating from the Boundary of an Internal Circular Hole," *J. Mathematics Physics,* Vol. *35* (1956), p. 60
4. J. M. Whitney and R. J. Nuismer, "Stress Fracture Criteria for Laminated Composites Containing Stress Concentrations," *J. Composite Materials,* Vol. *8* (1974), p. 253.
5. R. J. Nuismer and J. M. Whitney, Uniaxial Failure of Composite Laminates Containing Stress Concentrations," *Fracture Mechanics of Composites,* ASTM STP 593, Proceedings of

ASTM Symposium in High Modulus Fibers and Their Composites, Gaithersburg, Md., ASTM, Philadelphia, 1975.

6. R. F. Karlak, "Hole Effects in a Related Series of Symmetrical Laminates," Proceedings of Failure Modes in Composites III, American Society of Metals, Chicago, 1977.
7. S. G. Lekhnitskii, *Theory of Elasticity of an Anisotropic Elastic Body,* translated by P. Fern, Holden-Day Inc., San Francisco, 1963.
8. N. J. Pagano and R. B. Pipes, "The Influence of Stacking Sequence on Laminate Strength," *J. Composite Materials,* Vol. *5* (1971), p. 50
9. J. M. Whitney and R. Y. Kim, "Effect of Stacking Sequence on the Notched Strength of Composite Laminates," AFML-TR-76-177, 1976.
10. H. Leaderman, "Elastic and Creep Properties of Filamentous Materials and Other High Polymers," The Textile Foundation, Washington, 1943.
11. J. D. Ferry, *Viscoelastic Properties of Polymers,* John Wiley and Sons, Inc., New York, 1961.

A Micromechanical Prediction of Initial Yield Surfaces of Unidirectional Composites

K. WAKASHIMA, YOSHIO SUZUKI AND S. UMEKAWA

Research Laboratory of Precision Machinery and Electronics
Tokyo Institute of Technology
Nagatsuta, Midori-ku, Yokohama 227, Japan

(Received September 15, 1978)

ABSTRACT

For two-phase materials composed of ductile matrices reinforced with plastically nondeformable inclusions, the yield condition that has the same form as Tsai and Wu's phenomenological criterion has been deduced through a micromechanics analysis. Basic assumptions made are that the matrix material obeys the Levy-von Mises flow rule, and that matrix plastic strains are uniform. In general, the shape of the yield surface can be proved to be hyperellipsoidal.

A further analysis is devoted to the case where the inclusions can be treated as identical ellipsoids with corresponding axes aligned. Based on the results, numerical computation of yield strength parameters is carried out for unidirectional boron fiber-aluminum composites subjected to uniform temperature changes prior to external loading. It is shown that yielding cannot take place under certain combined triaxial (not hydrostatic) stresses. Some discussion related to the effect of prior temperature changes is also given.

SCOPE

FOR MATERIALS STRENGTH characterization and design purposes, various phenomenological failure criteria for composite materials have been proposed [1]. The most comprehensive of the existing criteria is a tensor polynomial failure criterion proposed by Tsai and Wu [2]. Specific advantages of this criterion include (1) invariance under rotation or redefinition of coordinates; (2) transformation via known tensor transformation laws; and (3) symmetry properties similar to those of the stiffnesses and compliances. It also accounts for strength difference in tension and compression. Accordingly, the Tsai-Wu criterion is operationally simple in application to practical problems, and its validity has been demonstrated in some composite systems [1, 3].

Notwithstanding their technological importance, phenomenological failure criteria are hypotheses made arbitrarily, and so is the Tsai-Wu criterion. It is desirable to give analytic foundation to the phenomenological hypotheses through mechanistic approaches; but unfortunately, the task appears to be generally intractable because failures of composites are very complicated by a number of independent and interacting mechanisms such as matrix yielding and crack propagation, fiber breaks and microbuckling, interface decohesion, and so on. In fact, all phenomenological failure criteria in scalar form have nothing to do with the prediction of failure mechanisms. However, the problem becomes relatively simple to solve if we confine our attention to a particular mechanism of failure. On this line, Chou, McNamee and Chou [4] have demonstrated recently that for laminated media consisting of plastically isotropic layers of different materials, the Tsai-Wu criterion is a direct extension of the yield condition of von Mises provided yielding in the individual layers is the only mechanism of composite failure. In their study, however, the composite is assumed to be stress-free before external loading. Consequently, the derived expression lacks linear terms in stresses which generally exist in the Tsai-Wu criterion.

This paper is concerned with the prediction of the yield surface of a two-phase material in which a ductile matrix obeying the distortion energy criterion of yielding and the Levy-von Mises theory of plastic flow is directionally reinforced with plastically nondeformable inclusions. The main objective is to show the correlation with the Tsai-Wu criterion. We will make much account of the effect of initial internal stresses due to thermal expansion mismatch between the constituents; this effect is of particular importance in metallic matrix composites since most of them are consolidated at high temperatures. Theoretically, our analysis is based on established principles pertinent to a certain potential energy function. Further, Eshelby's theory [5, 6] on an ellipsoidal inclusion is used as a convenient tool for approximate evaluating this potential energy function. The result predicted will be illustrated with some numerical examples for polycrystalline aluminum unidirectionally reinforced with continuous boron fibers.

CONCLUSIONS AND SIGNIFICANCE

A derived expression for the yield condition of the reinforced solid is shown to have the same form as the Tsai-Wu criterion, which consists of both quadratic and linear terms in stress components. Yield strength parameters, or "strength tensors" in Tsai and Wu's terminology, are given as functions of some coefficients involved in a potential energy function pertinent to micromechanics problems of this sort. On the basis of these functional expressions for the yield strength parameters, it can be proved that the composite yield surface constructed in a stress space is hyperellipsoidal in shape.

Special attention is directed to the case where the reinforcing phase is idealized as aligned, identical ellipsoids that are distributed in the matrix in a statistically uniform manner. Once this idealization is adopted, we are able to estimate the above micromechanical coefficients from knowledges of the aspect ratios of the inclusions as well as constituent elastic properties and volume fractions following Eshelby's [5, 6] elegant treatment of related inclusion problems. Interestingly, when applied to the prediction of composite elastic moduli, our approach can derive the same expressions as the best lower bounds [9, 11, 12] available at present.

Numerical results for unidirectional B-Al composites illustrate that under certain combined triaxial stresses yielding cannot take place, that is, the yield surface constructed in this particular three-dimensional stress space is open-ended to form an elliptic cylinder. In contrast with the case of plastically isotropic materials, the axis of the cylinder does not coincide with the hydrostatic stress axis, but varies with the volume content of boron fibers. Also illustrated is the fact that in most B-Al composites consolidated at high temperatures yielding can take place upon cooling to room temperature. In view of this fact, it is subsequent yielding rather than initial yielding that may be of practical importance. A full detail of subsequent yielding is not given in this paper, but it is suggested that the yield strength parameters for such composites may have different values that depend on the temperature path experienced.

SUPPLEMENTARY INFORMATION

Formulation

Consider a composite specimen (of unit volume) that has experienced a uniform temperature drop $\Delta T \equiv T - T_o$ from a reference temperature T_o prior to external loading and is in a state of internal stress due to the mismatch $\Delta\alpha$ in linear thermal expansion coefficients between a matrix and inclusions. Under applied triaxial stresses σ_i, the matrix alone is assumed to undergo uniform plastic strains ϵ_i^p. (For the time being, the usual contracted notation is used for stresses and strains; the range on subscripts is from 1 to 6 unless otherwise stated.)

A potential energy function Φ is now defined by the work stored in the specimen in the form of elastic strain energy, W^e, minus the work done by the external loading mechanism, W^{ext}. For small deformations, Φ can be written in the form,

$$\Phi = -\frac{1}{2}S_{ij}\sigma_i\sigma_j + \frac{1}{2}L_{ij}\epsilon_i^p\epsilon_j^p + L_i\epsilon_i^p\Delta\alpha\Delta T + \frac{1}{2}L\,(\Delta\alpha\Delta T)^2 - P_{ij}\sigma_i\epsilon_j^p \qquad (1)$$

where the usual summation convention applies. The coefficients S_{ij}, L_{ij}, L_i, L and P_{ij} are functions of constituent elastic properties and geometrical factors. When the inclusions can be treated as identical ellipsoids with corresponding axes aligned, it is possible, as will be seen later, to estimate all of these coefficients. Introducing proportionality factors σ and ϵ_i^p for σ_i and ϵ^p, respectively such that

$$\sigma_i = \tau_i \sigma \tag{2}$$

$$\epsilon_i^p = \lambda_i \epsilon^p \tag{3}$$

we have

$$\Phi = -\frac{1}{2} S_{ij}\tau_i\tau_j\sigma^2 + \frac{1}{2} L_{ij}\lambda_i\lambda_j (\epsilon^p)^2 + L_i\lambda_i\epsilon^p \Delta\alpha\Delta T + \frac{1}{2} L (\Delta\alpha\Delta T)^2 - P_{ij}\tau_i\lambda_j\sigma\epsilon^p \tag{4}$$

We now consider the virtual plastic deformation characterized by the variation $\delta\epsilon^p$; then the principle of virtual work states that

$$\delta W^e + \delta W^p = \delta W^{ext} \tag{5}$$

where δW^p is the plastic work dissipated. Assuming the matrix material to obey the distortion energy criterion of yielding and the Levy-von Mises theory of plastic flow, we have

$$\delta W^p = (1 - f)\, Y_o (\sqrt{2/3}\,\lambda)\, |\delta\epsilon^p| \tag{6}$$

with

$$\lambda \equiv \{\lambda_1^2 + \lambda_2^2 + \lambda_3^2 + \frac{1}{2}(\lambda_4^2 + \lambda_5^2 + \lambda_6^2)\}^{1/2} \tag{7}$$

where Y_o is the uniaxial tensile yield strength, and $(1 - f)$ is the fractional volume of the matrix. Substituting Equations (4) and (6) into Equation (5) leads to

$$\sigma = \frac{L_{ij}\lambda_i\lambda_j\epsilon^p + L_i\lambda_i\Delta\alpha\Delta T + k\lambda}{P_{ij}\tau_i\lambda_j} \tag{8}$$

with

$$k \equiv \pm\sqrt{2/3}\,(1 - f)\, Y_o \tag{9}$$

where the positive sign refers to the forward straining ($\delta\epsilon^p > 0$), and the negative sign to the backward straining ($\delta\epsilon^p < 0$). Denoting by σ_o the value of σ at which yielding starts, we have

$$\sigma_o \equiv \sigma_{(\epsilon^p \to 0)} = \frac{L_i\lambda_i\Delta\alpha\Delta T + k\lambda}{P_{ij}\tau_i\lambda_j} \tag{10}$$

Here, λ_i are still unknowns, and therefore σ_o is not definitive. However, a unique, energetically feasible value σ_o^* can be obtained through optimization of the potential energy function $\Phi_o \equiv \Phi_{(\epsilon^p \to 0)}$. From $\partial\Phi_o/\partial\lambda_m = 0$, it follows that

$$L_m \Delta\alpha\Delta T + k\,(\partial\lambda/\partial\lambda_m) - P_{im}\tau_i\sigma_o^* = 0 \tag{11}$$

where

$$\partial\lambda/\partial\lambda_m = \begin{cases} \lambda_m/\lambda & (m = 1, 2, 3) \\ \lambda_m/(2\lambda) & (m = 4, 5, 6) \end{cases} \tag{12}$$

On the other hand, from the constancy-of-volume relationship $\epsilon_1^p + \epsilon_2^p + \epsilon_3^p = 0$, not all λ_i are independent but

$$\lambda_1 + \lambda_2 + \lambda_3 = 0 \tag{13}$$

Therefore, using the method of Lagrange's multipliers, we have

$$\lambda_m = \begin{cases} (\lambda/k)\,(P_{im}\tau_i\sigma_o^* - L_m\Delta\alpha\Delta T - \beta) & (m = 1, 2, 3) \\ (2\lambda/k)\,(P_{im}\tau_i\sigma_o^* - L_m\Delta\alpha\Delta T) & (m = 4, 5, 6) \end{cases} \tag{14}$$

where β is a multiplier. Substitution of Equation (14) into Equation (13) yields

$$\beta = \frac{1}{3}(P_{i1} + P_{i2} + P_{i3})\,\tau_i\sigma_o^* - \frac{1}{3}(L_1 + L_2 + L_3)\,\Delta\alpha\Delta T \tag{15}$$

and hence,

$$\lambda_m = \begin{cases} (\lambda/k)\,(P'_{im}\tau_i\sigma_o^* - L'_m\Delta\alpha\Delta T) & (m = 1, 2, 3) \\ (2\lambda/k)\,(P'_{im}\tau_i\sigma_o^* - L'_m\Delta\alpha\Delta T) & (m = 4, 5, 6) \end{cases} \tag{16}$$

where

$$P'_{im} \equiv P_{im} - \frac{1}{3}(P_{i1} + P_{i2} + P_{i3})\,(\delta_{m1} + \delta_{m2} + \delta_{m3}) \tag{17}$$

$$L'_m \equiv L_m - \frac{1}{3}(L_1 + L_2 + L_3)\,(\delta_{m1} + \delta_{m2} + \delta_{m3}) \tag{18}$$

and δ_{ij} is the Kronecker delta. It therefore follows from Equations (7) and (16) that

$$(P'_{i1}\tau_i\sigma_o^* - L'_1\Delta\alpha\Delta T)^2 + (P'_{i2}\tau_i\sigma_o^* - L'_2\Delta\alpha\Delta T)^2$$
$$+ (P'_{i3}\tau_i\sigma_o^* - L'_3\Delta\alpha\Delta T)^2 + 2\,(P'_{i4}\tau_i\sigma_o^* - L'_4\Delta\alpha\Delta T)^2 \tag{19}$$
$$+ 2\,(P'_{i5}\tau_i\sigma_o^* - L'_5\Delta\alpha\Delta T)^2 + 2\,(P'_{i6}\tau_i\sigma_o^* - L'_6\Delta\alpha\Delta T)^2 = k^2$$

Denoting by σ_i^* the composite yield stresses (i.e., $\sigma_i^* = \tau_i\sigma_o^*$) and rearranging Equation (19) with respect to quadratic and linear terms in σ_i^*, we finally obtain

$$F_{ij}\sigma_i^*\sigma_j^* + F_i\sigma_i^* = 1 \tag{20}$$

where

$$F_{ij} = (1/K^2)\,[P'_{i1}P'_{j1} + P'_{i2}P'_{j2} + P'_{i3}P'_{j3} + 2\,(P'_{i4}P'_{j4} + P'_{i5}P'_{j5} + P'_{i6}P'_{j6})]$$
$$F_i = -(2\Delta\alpha\Delta T/K^2)\,[P'_{i1}L'_1 + P'_{i2}L'_2 + P'_{i3}L'_3 + 2\,(P'_{i4}L'_4 + P'_{i5}L'_5 + P'_{i6}L'_6)] \tag{21}$$

with

$$K^2 \equiv k^2 - (\Delta\alpha\Delta T)^2\,[L_1'^2 + L_2'^2 + L_3'^2 + 2\,(L_4'^2 + L_5'^2 + L_6'^2)]$$

Note that the above expression, Equation (20), is formally identical with a phenomenological failure criterion proposed by Tsai and Wu [2]. For the yield strength parameters F_{ij} given by Equation (21), it is an elementary matter to show that

$$F_{ii}F_{jj} - F_{ij}^2 \geqslant 0$$

where repeated subscripts are NOT summations. This inequality is called by Tsai and Wu a "stability condition," which ensures that the shape of the yield surface will be hyperellipsoidal.

Finally, it is added that for the yield surface to exist in a stress space K^2 in Equation (21) is required to take a positive value, but this requirement is not satisfied once $|\Delta T|$ exceeds a certain limit. Relating to this, some discussion will be given in a later section.

Estimates of Yield Strength Parameters

The final result in the preceding section shows that the yield surface of a particular composite can be theoretically constructed if its potential energy function Φ is given explicitly from a micromechanical point of view. In general, however, it seems difficult to obtain such an expression of Φ for arbitrary phase geometry. We therefore assume here that all inclusions can be treated as identical ellipsoids with corresponding axes aligned. This assumption enables us to use Eshelby's theory [5, 6] on the so-called transformation and inhomogeneity problems of an ellipsoidal inclusion. Note however that the straightforward application of his original theory is valid only to the case where the fractional volume f of the inclusions is small enough to ignore elastic interactions of the inclusions. Gathering suggestions from [7], an approximate method of modifying Eshelby's theory has been devised previously [8] to take the elastic interactions into account, and we will use this method for further development of our analysis. Then, it is only necessary to replace the original expression: $\epsilon_{ij}^C = S_{ijkl}\epsilon_{kl}^T$, for the relation between the constrained and stress-free strains ϵ_{ij}^C and ϵ_{ij}^T in the inclusion by

$$\epsilon_{ij}^C = (1-f)\, S_{ijkl}\epsilon_{kl}^T + f\epsilon_{ij}^T \tag{22}$$

(In this section the tensor notation is used mainly; the range on subscripts is from 1 to 3.) S_{ijkl}, so-called Eshelby's tensor, are dimensionless constants depending on the axial ratios of the ellipsoid (inclusion) and Poisson's ratio (of the matrix). Obviously, Equation (22) becomes identical with the original expression at the limit $f \to 0$. Furthermore, $\epsilon_{ij}^C \to \epsilon_{ij}^T$ when $f \to 1$; this also is physically acceptable because there is no cause for additional (elastic) strains to be induced. (For a new derivation of Equation (22), see Appendix.) Based upon the above modification, the potential energy function Φ can be expressed in terms of micromechanical parameters as [5, 6]

$$\Phi = -\frac{1}{2}C_{ijkl}^{**-1}\sigma_{ij}\sigma_{kl} + \frac{1}{2}L_{ijkl}\epsilon_{ij}^p\epsilon_{kl}^p - L_{iikl}\epsilon_{kl}^p\Delta\alpha\Delta T + \frac{1}{2}L_{iikk}(\Delta\alpha\Delta T)^2 - P_{ijkl}\sigma_{ij}\epsilon_{kl}^p \tag{23}$$

where

$$C_{ijkl}^{**-1} = [\delta_{ip}\delta_{jq} - fZ_{ijmn}^{-1}(C_{mnpq}^* - C_{mnpq})]\, C_{pqkl}^{-1} \tag{24}$$

$$L_{ijkl} = f(1-f)\, C_{ijmn}(\delta_{mp}\delta_{nq} - S_{mnpq})\, Z_{pqrs}^{-1}C_{rskl}^* \tag{25}$$

$$P_{ijkl} = \frac{1}{2}(\delta_{ik}\delta_{jl} + \delta_{il}\delta_{jk}) - fZ_{ijmn}^{-1}C_{mnkl}^* \tag{26}$$

with

$$Z_{ijkl} = (1-f)(C_{ijmn}^* - C_{ijmn})\, S_{mnkl} + (1-f)\, C_{ijkl} + fC_{ijkl}^* \tag{27}$$

(A tensor with the superscript -1 is defined such that $A^{-1}_{ijmn}A_{mnkl} = \delta_{ik}\delta_{jl}$.) C_{ijkl}, C^{*}_{ijkl} and C^{**}_{ijkl} are the elastic stiffness tensors of the matrix, inclusions and composite, respectively; and $\Delta\alpha \equiv \alpha^{*} - \alpha$, where α and α^{*} are the (isotropic) linear thermal expansion coefficients of the matrix and inclusions, respectively.

Though aside from our present subject, it is worthwhile to note that the composite elastic moduli estimated from Equation (24) coincide with some existing predictions. Assuming spherical inclusions, for example, we obtain the same expressions as Hashin and Shtrikman's [9] lower bounds for the overall bulk and shear moduli of a statistically homogeneous isotropic two-phase material (provided the inclusions are stiffer than the matrix); they are also equivalent to Kerner's [10] predictions. If, furthermore, the inclusions are treated as an extremely long prolate spheroid, the expressions for five independent moduli that are identical with the lower bounds given by Hill [11] and Hashin [12] for a statistically homogeneous and transversely isotropic two-phase material are reproduced. Incidentally, four of the five are known to be identical with Hashin and Rosen's [13] exact solutions for the "composite cylinder assemblage."

We now turn back to our present subject. From the symmetry properties of L_{ijkl} and P_{ijkl}, it can be easily shown that non-zero components of L_i and P_{ij} in Equation (1) are given as

$$
\begin{aligned}
L_1 &= -(L_{1111} + L_{2211} + L_{3311}) \\
L_2 &= -(L_{1122} + L_{2222} + L_{3322}) \\
L_3 &= -(L_{1133} + L_{2233} + L_{3333})
\end{aligned}
\tag{28}
$$

$$
\begin{aligned}
P_{ij} &= P_{iijj}\ (i, j = 1, 2, 3; \text{NOT summations for } i \text{ and } j) \\
P_{44} &= 2P_{2323} \\
P_{55} &= 2P_{3131} \\
P_{66} &= 2P_{1212}
\end{aligned}
\tag{29}
$$

provided the cartesian coordinate axes are taken parallel to the principal axes of the ellipsoid (inclusion). Then, as noted by Tsai and Wu for specially orthotropic materials, the number of independent yield strength parameters reduces from 6 to 3 for F_i and from 21 to 9 for F_{ij}. In the next section, we will be concerned with unidirectional filamentary composites with particular reference to polycrystalline aluminum reinforced with continuous boron fibers. Envisaging a typical fiber as the limiting case where the semi-major axis of a prolate spheroid, lying parallel to the x_3 coordinate axis, becomes infinite, we have non-zero components of S_{ijkl} as [5]

$$
\begin{aligned}
&S_{1111} = S_{2222} = \frac{5-4\nu}{8(1-\nu)} \\
&S_{1122} = S_{2211} = -\frac{1-4\nu}{8(1-\nu)} \\
&S_{1133} = S_{2233} = \frac{\nu}{2(1-\nu)} \\
&S_{1212} = S_{1221} = S_{2112} = S_{2121} = \frac{3-4\nu}{8(1-\nu)} \\
&S_{2323} = S_{2332} = S_{3223} = S_{3232} = S_{3131} = S_{3113} = S_{1331} = S_{1313} = \frac{1}{4}
\end{aligned}
\tag{30}
$$

where ν is Poisson's ratio of the matrix. The above knowledge makes it possible to estimate the yield strength parameters F_i and F_{ij} in Equation (20) and therefore to construct the yield surface of the composite in a six-dimensional stress space.

Numerical Examples and Discussion

In the B-Al family of composites, the yield strength Y_o of the matrix material may change considerably with chemical composition, heat treatment, and temperature. It is thus convenient to express the composite yield stresses in dimensionless terms of σ_i^*/Y_o; then the yield condition, Equation (20), can be rewritten as

$$Y_o^2 F_{ij}(\sigma_i^*/Y_o)(\sigma_j^*/Y_o) + Y_o F_i(\sigma_i^*/Y_o) = 1 \tag{31}$$

For numerical estimation of $Y_o^2 F_{ij}$ and $Y_o F_i$, the following values are assumed:

	Al-matrix	B-fiber
Young's modulus	71500 MPa	410000 MPa
Poisson's ratio	0.33	0.21

Some of the results obtained using a Hitac M-180 computer are summarized in Table 1. Graphical representations of these results are not given here, but the following conclusions may be easily drawn with the aid of established theories of solid geometry.

Table 1. Estimated Values of $Y_o^2 F_{ij}$ and $Y_o F_i$ for Unidirectional B-Al Composites. (The 3-axis is taken parallel to the filament axis; the 1–2 plane is thus the plane of transverse isotropy. The value represented by, say, −0.27254 + 01 should be read as $-0.27254 \times 10^{+01}$.)

Parameter*	Fiber Volume Fraction f				
	0.1	0.3	0.5	0.7	0.9
a	0.81000+00	0.49000+00	0.25000+00	0.90000-01	0.10000-01
b	0.50165+01	0.92720+01	0.62503+01	0.24827+01	0.28426+00
A	0.67220+00	0.32181+00	0.13740+00	0.42378-01	0.40943-02
B	0.37018+00	0.82070-01	0.21441-01	0.46656-02	0.34615-03
C	0.21121+01	0.99114+00	0.40367+00	0.11868+00	0.10973-01
D	0.22220+01	0.11365+01	0.49671+00	0.15489+00	0.15051-01
E	-0.20785+00	-0.55606-01	-0.16835-01	-0.41305-02	-0.33876-03
F	-0.43879+00	-0.24646+00	-0.11096+00	-0.35065-01	-0.34313-02
G	0.15303+01	0.11821+01	0.57488+00	0.19056+00	0.19415-01
H	-0.27254+01	-0.17447+01	-0.73216+00	-0.21525+00	-0.19839-01

*

$$Y_o^2 F_{ij} = \frac{1}{a - b\xi^2} \begin{bmatrix} A & F & E & 0 & 0 & 0 \\ & A & E & 0 & 0 & 0 \\ & & B & 0 & 0 & 0 \\ & & & C & 0 & 0 \\ & & & & C & 0 \\ \text{Sym.} & & & & & D \end{bmatrix} \qquad Y_o F_i = \frac{\xi}{a - b\xi^2} \begin{bmatrix} G \\ G \\ H \\ 0 \\ 0 \\ 0 \end{bmatrix}$$

$$\xi \equiv \frac{\Delta\alpha\Delta T}{Y_o} \times 10^4 \quad (Y_o \text{ in units of MPa})$$

First, yield surfaces constructed in three-dimensional (σ_i/Y_o) (σ_j/Y_o) (σ_k/Y_o) normalized stress spaces are ellipsoidal in shape; all but one are closed surfaces. The single exception is for the case: $i = 1$, $j = 2$ and $k = 3$, where one of the principal axes of the ellipsoid becomes infinite to form an elliptic cylinder. The axis of the cylinder lies in the plane: $(\sigma_1/Y_o) = (\sigma_2/Y_o)$, making an inclination angle θ ($0° < \theta < 54.7°$) with the (σ_3/Y_o) normalized stress axis; θ decreases with increasing f. Therefore, as has been emphasized by several authors [14, 15, 16], hydrostatic loading may cause yielding in these kinds of materials. Note however that there is a certain combination of normal stresses under which yielding cannot take place. Numerically, the stress ratios, $\sigma_1:\sigma_2:\sigma_3$, for this situation to arise are, for example, 1:1:1.36 when $f = 0.3$ and 1:1:1.77 when $f = 0.7$.

A second conclusion is concerned with the effect of temperature changes prior to external loading. When $\Delta T = 0$, linear terms in σ_i^*/Y_o all vanish from Equation (31) and thus obviously, all ellipsoidal yield surfaces constructed in the (σ_i/Y_o) (σ_j/Y_o) (σ_k/Y_o) spaces are symmetric with respect to the origin. As $|\Delta T|$ increases, these surfaces, in general, shift gradually in a certain direction of the respective spaces, keeping their size, shape, and orientation unchanged. For the yield surface in the (σ_1/Y_o) (σ_2/Y_o) (σ_3/Y_o) space, for example, this translation occurs in the positive or negative direction of the hydrostatic stress axis according as ΔT is negative or positive. However, as has been noted in a previous section, further increase in $|\Delta T|$ eventually causes the yield surfaces to disappear from the respective stress spaces. This means that yielding can take place even in the absence of externally applied stresses when the composite is heated or cooled to a certain extent from the reference temperature T_o. The magnitude of uniform temperature changes $|\Delta T|$ to cause such yielding is proportional to $Y_o/|\Delta\alpha|$ and is not very large in the B-Al system, e.g., $|\Delta T| \simeq 54$ and 45°C for annealed and recrystallized commercially pure aluminum matrices ($Y_o \simeq 35$ MPa) reinforced with 30 and 70 v/o boron fibers, respectively. The reference temperature T_o, which is defined as a stress-free temperature, may now be taken almost equal to the temperature of hot-pressing, typically 450 ∿ 500°C. Consequently, in accordance with previous experimental observations [17], considerable amounts of plastic strains can be introduced into the Al matrices upon cooling to room temperature. In view of this fact, subsequent yielding rather than initial yielding will be of practical importance. The analysis developed in this study can be easily extended to the case of subsequent yielding by dividing matrix plastic strains into pre-loading and post-loading parts. Although we will not enter here into analytical details, it can be shown that the pre-loading part of matrix plastic strains depends not simply on the magnitude of ΔT but on the temperature path experienced, and consequently so is the state of internal stress before external loading. As a result, the yield strength parameters F_i and F_{ij} pertinent to subsequent yielding are dependent on the thermal history. For a related experimental study, see [18].

Finally, it is added that for yielding under uniaxial normal stresses applied in the fiber direction, our numerical results (Table 1) are almost the same as those predicted from the "rule-of-mixtures" equation.

NOMENCLATURE

σ_i (σ_{ij})	applied stresses
ϵ_i^p (ϵ_{ij}^p)	uniform plastic strains in matrix
σ, ϵ^p	proportionality factors for σ_i and ϵ^p, respectively
τ_i	$\equiv \sigma_i/\sigma$
λ_i	$\equiv \epsilon_i^p/\epsilon^p$
σ_i^*	yield stresses of composite
ΔT	temperature change from a stress-free temperature T_o
$\Delta\alpha$	difference in linear thermal expansion coefficients of constituents
f	fractional volume of reinforcement (inclusion)
W^e, W^p, W^{ext}	work stored, work dissipated, and work done, respectively
Φ	potential energy function defined as $W^e - W^{ext}$
L, L_i, L_{ij}, P_{ij}, S_{ij}	constants dependent on constituent elastic properties and phase geometry

Y_o uniaxial tensile yield strength of matrix
F_{ij}, F_i yield strength parameters of composite

REFERENCES

1. E. M. Wu, "Phenomenological Anisotropic Failure Criterion," in *Composite Materials*, L. J. Broutman and R. H. Krock, Eds., Vol. 2, p. 353, Academic Press, 1974.
2. S. W. Tsai and E. M. Wu, "A General Theory of Strength for Anisotropic Materials," *J. Composite Materials*, Vol. 5 (1971), p. 58.
3. E. M. Wu, "Optimal Experimental Measurements of Anisotropic Failure Tensors," *J. Composite Materials*, Vol. 6 (1972), p. 472.
4. P. C. Chou, B. M. McNamee and D. K. Chou, "The Yield Criterion of Laminated Media," *J. Composite Materials*, Vol. 7 (1973), p. 22.
5. J. D. Eshelby, "The Determination of the Elastic Field of an Ellipsoidal Inclusion, and Related Problems," *Proc. Roy. Soc.* (London), Vol. 241A (1957), p. 376.
6. J. D. Eshelby, "The Elastic Field Outside an Ellipsoidal Inclusion," *Proc. Roy. Soc.* (London), Vol. 252A (1959), p. 561.
7. T. Mori and K. Tanaka, "Average Stress in Matrix and Average Elastic Energy of Materials with Misfitting Inclusions," *Acta Metall.*, Vol. 21 (1973), p. 571.
8. K. Wakashima, M. Otsuka and S. Umekawa, "Thermal Expansions of Heterogeneous Solids Containing Aligned Ellipsoidal Inclusions," *J. Composite Materials*, Vol. 8 (1974), p. 391.
9. Z. Hashin and S. Shtrikman, "A Variational Approach to the Theory of Multiphase Materials," *J. Mech. Phys. Solids*, Vol. 11 (1963), p. 127.
10. E. H. Kerner, "The Elastic and Thermoelastic Properties of Composite Media," *Proc. Phys. Soc.*, Vol. 69B (1956), p. 808.
11. R. Hill, "Theory of Mechanical Properties of Fibre-Strengthened Materials: I. Elastic Behaviour," *J. Mech. Phys. Solids*, Vol. 12 (1964), p. 199.
12. Z. Hashin, "On Elastic Behaviour of Fibre Reinforced Materials of Arbitrary Transverse Phase Geometry," *J. Mech. Phys. Solid*, Vol. 13 (1965), p. 119.
13. Z. Hashin and B. W. Rosen, "The Elastic Moduli of Fiber Reinforced Materials," *J. Appl. Mech.*, Vol. 31 (1964), p. 223.
14. T. H. Lin, D. Salinas and Y. M. Ito, "Effects of Hydrostatic Stress on the Yielding of Cold Rolled Metals and Fiber-Reinforced Composites," *J. Composite Materials*, Vol. 6 (1972), p. 409.
15. G. J. Dvorak, M. S. M. Rao and J. Q. Tarn, "Yielding in Unidirectional Composites Under External Loads and Temperature Changes," *J. Composite Materials*, Vol. 7 (1973), p. 194.
16. K. Wakashima and S. Umekawa, "Fiber-Composite Response to Hydrostatic States of Stress," *Met. Trans.*, Vol. 7A (1976), p. 1952.
17. K. G. Kreider and V. M. Patarini, "Thermal Expansion of Boron Fiber-Aluminum Composites," *Met. Trans.*, Vol. 1 (1970), p. 3431.
18. K. Wakashima, T. Kawakubo and S. Umekawa, "The Effect of Thermal History on the Axial Tensile Behavior of an Aligned Fiber Composite," *Met. Trans.*, Vol. 6A (1975), p. 1755.

APPENDIX

This Appendix is devoted to a further clarification of our micromechanics analysis which is based on the utility of Equation (22).

Eshelby [5] dealt with the following transformation problem: An ellipsoidal region (the "inclusion") in an infinite homogeneous isotropic elastic medium with the stiffness tensor C_{ijkl} undergoes stress-free (transformation) strains ϵ_{ij}^T which, but for the constraint imposed by its surroundings (the "matrix"), would be arbitrary homogeneous strains. What is the elastic state of inclusion and matrix? His solutions to the constrained strains ϵ_{ij}^C and stresses σ_{ij}^I in the inclusion are such that

$$\epsilon_{ij}^C = S_{ijkl}\epsilon_{kl}^T \tag{A1}$$

$$\sigma_{ij}^I = C_{ijkl}\,(\epsilon_{kl}^C - \epsilon_{kl}^T) \tag{A2}$$

where S_{ijkl} are dimensionless constants ("Eshelby's tensor") depending on the axial ratios of the inclusion and Poisson's ratio of the medium. In this Appendix, we attempt to obtain solutions corresponding to Equations (A1) and (A2) for the case where a large number of identical ellipsoidal inclusions, all being aligned, are so distributed in the medium as to produce a random and statistically homogeneous dispersion. Under the circumstances, it is clear that detailed field analyses are in general impossible; all we may know for field values is at best their statistical average.

First, the individual inclusions are labeled as V^1, V^2, V^3, ... Let any one of them, V^1, say, be fixed at the origin as shown in Figure A.1, and let us consider the probability p that any given point $X \equiv (x_1^*, x_2^*, x_3^*)$ outside V^1 is occupied by any one of the remainder V^n $(n = 2, 3, \ldots)$. If the point X is included within some one of V^n, V^i, say, the center of V^i should lie inside an ellipsoid E_1 of the form,

$$(x_1 - x_1^*)^2/a^2 + (x_2 - x_2^*)^2/b^2 + (x_3 - x_3^*)^2/c^2 = 1 \tag{A3}$$

where *a, b* and *c* are the principal semi-axes of ellipsoidal inclusions. On the other hand, this V^i must not overlap with the fixed V^1, so that, furthermore, the center of V^i should lie outside an ellipsoid E_2 of the form,

$$x_1^2/(2a)^2 + x_2^2/(2b)^2 + x_3^2/(2c)^2 = 1 \tag{A4}$$

Since the volume fraction f of the inclusions may be interpreted as the probability of finding an inclusion at an arbitrary point in the medium *when none of the inclusions is fixed*, the probability concerned, p, may be given as

$$p = \frac{\Omega}{\Omega^\circ} f \tag{A5}$$

where Ω° is the domain volume inside E_1, and Ω is the domain volume inside E_1 and outside E_2.

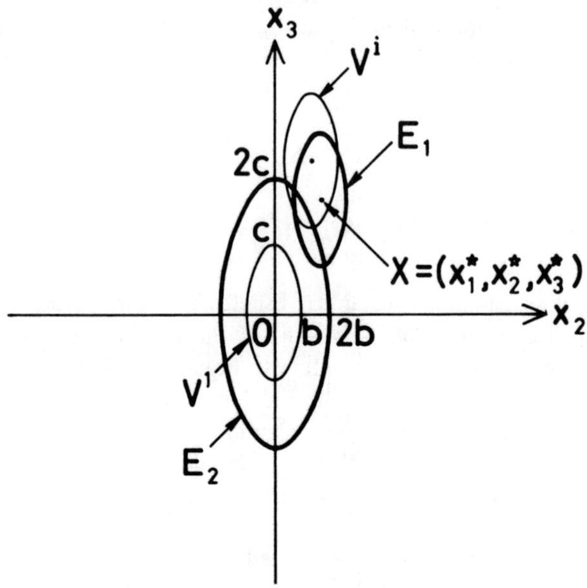

Figure A1. *Geometry used in evaluating the probability p that any given point X outside a fixed invlusion V^1 is occupied by any one of the surrounding inclusions, V^i, say.*

In order to determine iso-p contours in the medium, we introduce the y-space such that $y_1 = x_1/a$, $y_2 = x_2/b$ and $y_3 = x_3/c$. Then the point X is transformed into the point $Y \equiv (y_1^*, y_2^*, y_3^*)$, where $y_1^* = x_1^*/a$, $y_2^* = x_2^*/b$ and $y_3^* = x_3^*/c$; and the ellipsoids E_1 and E_2 into spheres S_1 and S_2 of the form,

$$S_1 : \ (y_1 - y_1^*)^2 + (y_2 - y_2^*)^2 + (y_3 - y_3^*)^2 = 1 \tag{A6}$$

$$S_2 : \ y_1^2 + y_2^2 + y_3^2 = 4 \tag{A7}$$

Noting that the volume element in the x-space is linearly proportional to that in the y-space, we have

$$\Omega/\Omega^\circ = \omega/\omega^\circ \tag{A8}$$

where ω° is the domain volume inside S_1, and ω is the domain volume inside S_1 and outside S_2. Thus, it immediately follows from consideration of the geometrical symmetry that the iso-p contours in the y-space consist of a set of concentric spherical surfaces of the form

$$y_1^{*2} + y_2^{*2} + y_3^{*2} = [F(p)]^2 \tag{A9}$$

Here, $F(p)$ is a certain function of p, satisfying that $F(p) = 1$ when $p = 0$, $3 > F(p) > 1$ when $f > p > 0$, and $F(p) \geqslant 3$ when $p = f$. Therefore, the iso-p contours in the x-space are expressed as

$$x_1^{*2}/a^2 + x_2^{*2}/b^2 + x_3^{*2}/c^2 = [F(p)]^2 \tag{A10}$$

In the present problem all inclusions are considered to have identical stress-free strains ϵ_{ij}^T, and consequently the statistical distribution of stress-free strains around the arbitrarily fixed inclusion V^1 may be given simply as $\epsilon_{ij}^T p$.

Based on the above result, we next analyze stresses and strains produced in V^1. Consider first how much constrained strains would be created in V^1 if a small amount of stress-free strains, $\epsilon_{ij}^T \Delta p$, were distributed uniformly throughout the region *outside* an arbitrarily chosen (ellipsoidal) iso-$\epsilon_{ij}^T p$ contour. This solution can be easily obtained through direct application of Eshelby's solution after superimposing $-\epsilon_{ij}^T \Delta p$ throughout the medium:

$$\Delta\epsilon_{ij}^C = -S_{ijkl}\epsilon_{ij}^T \Delta p + \epsilon_{ij}^T \Delta p \tag{A11}$$

Here, it is essential to notice that S_{ijkl} in Equation (A11) are (dimensionless) constants depending on the axial ratios of the (arbitrarily chosen) ellipsoidal iso-$\epsilon_{ij}^T p$ contour, and hence, exactly equal to S_{ijkl} in Equation (A1). Keeping this point in mind and using the superposition principle in the theory of elasticity, we can obtain the constrained strains created in V^1 due to the presence of surrounding inclusions, that is,

$$-S_{ijkl}\epsilon_{kl}^T \int_0^f dp + \epsilon_{ij}^T \int_0^f dp = -f S_{ijkl}\epsilon_{kl}^T + f\epsilon_{ij}^T$$

Consequently, the solution corresponding to Equation (A1) for our multiple inclusion problem is given by the sum of the above result and Equation (A1):

$$\epsilon_{ij}^C = (1-f)\, S_{ijkl}\epsilon_{kl}^T + f\epsilon_{ij}^T$$

The inclusion stresses σ_{ij}^I are, of course, expressed in the same form as Equation (A2).

Thus far, we have considered the transformation problem for a homogeneous material. Since the derived expressions for ϵ_{ij}^C and σ_{ij}^I are independent of position, that is, all inclusions are in the same homogeneous elastic state, we may also deal, following Eshelby's elegant treatment, with the transformation and inhomogeneity problems for that material in which ellipsoidal inclusions have elastic constants differing from those of the matrix.

Predicting the Time-Temperature Dependent Axial Failure of B/A1 Composites

JAMES A. DICARLO, PHYSICIST

National Aeronautics and Space Administration
Lewis Research Center
Cleveland, Ohio 44135

(Received May 5, 1980)

ABSTRACT

Experimental and theoretical studies were conducted in order to understand and predict the effects of time, temperature, and stress on the axial failure modes of boron fibers and B/A1 composites. Due to the anelastic nature of boron fiber deformation, it was possible to determine simple creep functions which can be employed to accurately describe creep and fracture stress of as-produced fibers. Analysis of damping and strength data for B/6061 A1 composites indicates that fiber creep and the effects of creep on fiber fracture are measurably reduced by the composite fabrication process. The creep function appropriate for fibers within B/A1 composites was also determined. A fracture theory is presented for predicting the time-temperature dependence of the axial tensile strength for metal matrix composites in general and B/A1 composites in particular.

INTRODUCTION

THE ABILITY TO predict the failure modes of structural materials under various environmental conditions has both a practical and fundamental significance. On the practical side, the predictive theory and associated equations will allow the design engineer to estimate structural failure for conditions not covered in available test data. On the fundamental side, verification of the predictive theory by comparison with experimental data will confirm the mechanistic models employed to derive the theory. Such confirmation not only may allow future theoretical modifications which yield a more accurate predictive ability but also may lead to the development of practical techniques for improving structural performance.

The objective of this paper is to demonstrate that predictive equations can be developed that adequately describe the effects of time, temperature, and stress on the axial failure modes of B/Al composites. For many metal matrix composite systems the reinforcing fibers deform elastically so that time-temperature effects arise mainly from the mechanical properties of the matrix. However, for B/Al composites, both the matrix and fiber make major contributions to the time and temperature dependence of composite failure. This is due to the fact that in contrast to other ceramic fibers such as silicon carbide and alumina, boron fibers display a low temperature creep which has a significant effect on fiber fracture. For this reason the approach taken in this paper will be to first investigate boron fiber creep and fracture data in order to establish a model and equations that accurately describe the axial failure modes of as-produced fibers. Once this is accomplished, composite data will be examined to determine how these equations might be modified to describe fiber creep and fracture within B/6061 Al composites. As part of this examination, a general metal matrix composite fracture theory will be developed based on the primary fiber and matrix mechanisms which contribute to time and temperature-dependent axial composite failure.

PROCEDURE

Background

The low temperature deformation of a boron fiber has been observed to be characteristically anelastic [1,2,3]. That is, in a creep test upon application of a constant tensile stress σ at time $t = 0$, the total strain ε in the fiber increases with time t and temperature T according to

$$\varepsilon(t,T,\sigma) = \varepsilon_e(T,\sigma) + \varepsilon_a(t,T,\sigma) \qquad (1)$$

Here $\varepsilon_e = \sigma/E(T)$ is the time-independent elastic strain which depends on T only through the elastic Young's modulus E. The anelastic creep strain ε_a is zero at $t = 0$ but increases with time, temperature, and stress. No evidence of plastic strain or strains other than anelastic has been found in boron fiber deformation for temperatures up to 800 °C. Three properties which characterize ε_a and distinguish it from plastic strain are:

I. Linearity: ε_a is directly proportional to σ.

II. Equilibrium: After passage of sufficient time, ε_a reaches or relaxes to a unique equilibrium value.

III. Recoverability: Upon removal of σ, the developed ε_a completely disappears at a rate which is time and temperature dependent.

Because of property *I* it is convenient to introduce a stress-independent anelastic strain function *A* defined by

$$A(t,T) = (\varepsilon/\varepsilon_e) = 1 + (\varepsilon_a/\varepsilon_e) \,. \tag{2}$$

$A = 1$ for pure elastic behavior, and $A > 1$ for anelastic behavior. With this definition for *A*, Eq. (1) can be written as

$$\varepsilon(t,T,\sigma) = [\sigma/E(T)]A(t,T) \tag{3}$$

which is simply Hooke's law multiplied by the time-temperature dependent factor *A*. Clearly, to understand and predict the effects of anelastic deformation on the failure modes of boron fibers and B/A1 composites, one must have accurate knowledge of the boron *A* function. In this paper the results of various deformation experiments will be presented in which Eqs. (2) and (3) were employed to calculate the *A* functions for as-produced boron fibers and for fibers within B/A1 composites.

Deformation tests

The primary experiments employed to determine the *A* function for as-produced boron fibers were low-stress flexural stress relaxation (FSR) and flexural internal friction (FIF) tests performed at various temperatures up to 800 °C. Details of the test apparatus and applicable deformation theories are described elsehwere [1]. For the FSR tests, anelastic creep strains were allowed to develop for one hour, whereas for the FIF tests, creep occurred only in a time span of one vibration period which was typically of the order of one millisecond. This large difference in test time coupled with anelasticity theory permitted accurate extrapolation of the *A* function for time-temperature conditions not covered by the tests.

As predicted by anelasticity property *I*, the *A* functions measured by the low-stress flexural tests were found to be independent of the applied stress. However, when the FSR tests were conducted at stress levels above 50 ksi, (0.3 GN/m^2) an unexpected stress dependence for the *A* function was observed. To study this effect in greater detail, two types of high stress tensile experiments were performed: room temperature elongation experiments [4] at stress levels of about 400 ksi (2.8 GN/m^2), and stress rupture experiments on etched boron fibers (1) at temperatures from 200 to 1000 °C and at stress levels between 200 and 800 ksi (1.4 and 5.6 GN/m^2). Since the latter experiments have a direct bearing on a predictive fracture model for boron fibers, the details of the deformation theory involved in its interpretation will be discussed here.

After slightly etching 203 μm boron on tungsten fibers, Smith [5] observed that essentially all cases of fiber fracture could be explained by crack initiation within the region of the tungsten boride core. This result suggests a "composite fiber" fracture model in which an etched boron on tungsten fiber fractures whenever the total axial strain of the boron anelastic sheath becomes equal to the core fraction strain. By assuming a brittle elastic core with a fracture strain independent of time and temperature, one can then use Eq. (3) to express this model in the following form:

$$\bar{\varepsilon}_u = [\bar{\sigma}_u(t,T)/E(T)]A(t,T) = \text{CONSTANT} \tag{4}$$

Here $\bar{\varepsilon}_u$ is the average (ultimate) fracture strain of the etched fiber, and $\bar{\sigma}_u$ is the average (ultimate) fracture stress required to obtain $\bar{\varepsilon}_u$. In the stress rupture tests, $\bar{\sigma}_u(t,T)$ were measured for various times and temperatures. These data plus room temperature $\bar{\varepsilon}_u$ data were then inserted into Eq. (4) to determine the A function at the $\bar{\sigma}_u$ stress level.

To determine the A function for fibers within B/Al composites, it was necessary to perform a composite test in which essentially all deformation is fiber controlled. One test that fulfills this requirement is the internal friction or damping test conducted on B/Al composites which have been annealed near 400 °C. The experimental and theoretical details of this test are described elsewhere [6]. The annealing treatment essentially eliminates dislocation damping within the aluminum matrix, leaving the fibers as the only source of composite damping [7]. Application of the rule-of-mixtures to axial temperature-dependent damping data for unidirectional B/Al composites allowed accurate calculations of fiber damping as a function of temperature. As previously described for the FIF data [1], these fiber damping results were then used as deformation data to calculate the A function for fibers within B/Al composites. The stress levels of the composite damping test varied between 0.01 and 1 ksi (0.07 and 7 MN/m^2). In this range no stress effects on fiber damping were observed, indicating a true stress independence for the A function up to at least 1 ksi.

Specimens

The primary specimens employed for the single-fiber tests were 203 μm (8 mil) and 142 μm (5.6 mil) boron on tungsten fibers supplied by Avco Specialty Materials Division. During the chemical vapor deposition of the boron sheath, the original 13 μm tungsten substrate became completely borided to form a 17 μm diameter core region.

Because fiber coatings may have an affect on the fiber anelasticity, 142 μm boron fibers coated with a 1.5 μm thick silicon carbide layer were also examined. These fibers which also contained the 17 μm tungsten-boride core were supplied by Composite Technology Inc. under the tradename "Borsic".

The unidirectional B/A1 composite specimens employed for the damping experiments contained normally 50 volume percent boron or Borsic fiber in a 6061 A1 matrix. The specimens reinforced by 203 μm boron fibers were fabricated by TRW whereas those reinforced by the Borsic fibers were fabricated by Avco. Typical fabrication techniques involving diffusion bonding near 500 °C were employed for both composite types.

RESULTS AND DISCUSSION

As-Produced Fibers

Fiber-Creep — In previous work (1) it was determined that anelastic creep in boron fibers is a thermally-activated process. As such, the time and temperature conditions required to produce a certain anelastic strain ε_a are not independent variables. That is, fixing the test temperature fixes the deformation time at which the given strain will be reached. The results of the low-stress FSR and FIF test indicated that for boron fibers, the relationship between the time and temperature variables was best expressed in terms of a q parameter given by

$$q = (\ln t + 33.7)\, T(10^{-3}). \qquad (5)$$

Here t is deformation time in seconds and T is test temperature in degrees Kelvin. Thus, ε_a and the anelastic function A depend only on the one variable q rather than the two variables t and T. This result greatly simplified the experimental procedures required to determine the A function. For example, since q is weakly dependent on time ($\ln t$) and directly proportional to temperature, deformation tests were typically conducted by holding the deformation time constant and measuring the development of anelastic creep strain as a function of test temperature. The deformation strains and Equations (2) and (3) were then used to calculate $A(q)$ at the q value corresponding to the particular time-temperature test conditions.

The $A(q)$ results from the low-stress flexural tests on as-produced fibers are shown as curve A_L in Figure 1. The subscript L refers to the fact that the measurements were made at low stresses below 50 ksi (0.3 GN/m²) where $A(q)$ was observed to be stress independent. Actual data points are not indicated because they have negligible error and were measured almost continuously ($\Delta q \simeq 0.3\ K$). To put these results in perspective, a 60 sec test at room temperature corresponds to a q value of 28 K. Since $A_L \simeq 1$ for q values less than 15 K, it follows that at low stresses and short times, boron fibers deform essentially elastically at room temperature and below. However, at longer times or higher temperatures, these fibers will display anelastic creep, even at very low stress.

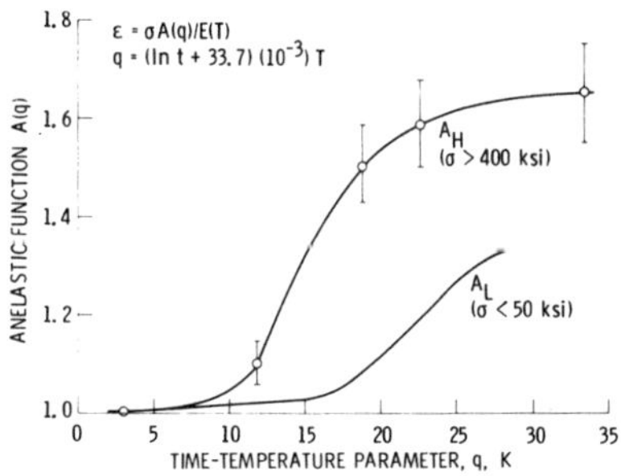

Figure 1. The anelastic strain functions for as-produced boron fibers.

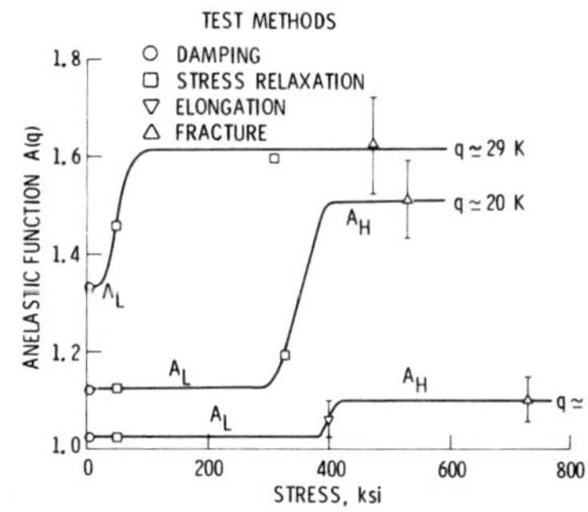

Figure 2. The effects of stress on the anelastic strain function for as-produced boron fibers.

As previously discussed, raising the stress level above 50 ksi (0.3 GN/m^2) produced an unexpected increase in the A function. To study this effect, tensile elongation and fracture tests were conducted on single fibers. In Figure 2 the A function results from the tensile and flexural tests are plotted as a function of stress for q values of 11, 20, and 29 K. For tests of one minute duration these values roughly correspond to test temperatures of 20°, 250°, and 500°C. Although the stress effect data are limited, the Figure 2 curves were drawn assuming a discontinuous behavior for the boron fiber A function. That is, as stress increases, the A function remains constant at the A_L level until at some transition stress σ^* where it rather abruptly increases to a constant A_H level as measured by the tensile fracture tests on etched fibers. The subscript H refers to the "high" stress level (>400 ksi (2.8 GN/m^2)) of these measurements (1). The q dependence of the A_H function is shown in Figure 1. The large error bars for the A_H data points are due primarily to a coefficient of variability of ~5 percent in the fracture stress data.

Although it is not obvious that the Figure 2 stress effect data support the assumption of an abrupt increase in A from one level to another, there does exist some indirect evidence for such behavior. For example, in previous work it was shown that boron fiber anelasticity could be explained by grain boundary type sliding of small substructural boron units [2,4]. Based on this microstructural model, the maximum A function to be expected for boron fibers is given approximately by the A_H curve of Figure 1. The fact that A_L is less that A_H suggests that at low stress all boron units cannot participate in creep, due possibly to the existence of some unknown internal "locking" mechanism. The stress-induced increase from A_L to A_H indicates that high stress can unlock the immobilized units, giving rise to maximum anelastic creep. If the locking mechanisms are all of one type, one might expect that the unlocking should occur over a narrow stress range. Experimental support for this may be found in the boron fiber torsional damping data or Firle [8] who observed that as shear stress is increased, fiber damping remains constant until some higher shear stress level at which it increases rather abruptly over a small stress range. Thus, the assumption is made that A_L and A_H are constant over certain stress regimes and that the transition from A_L to A_H occurs abruptly at σ^*. As indicated by the Figure 2 curves, σ^* decreases with increasing q or temperature, suggesting that the unlocking mechanism is thermally activated.

Summarizing the practical aspects of the above results, one can now predict creep of as-produced boron fibers by employing the following equation:

$$\varepsilon(t, T, \sigma) = [\sigma/E(T)]A_\sigma(q) \tag{6}$$

where $A_\sigma = A_L$ for $\sigma < \sigma^*$ and $A_\sigma = A_H$ for $\sigma > \sigma^*$. The q parameter is given by Equation (5) and $\sigma^*(q)$ can be estimated from the Figure 2 results. Accurate data for $E(T)$ were measured during the single fiber damping tests (1,4). These data are shown in Figure 3 in terms of the ratio $E/E(20\,°C)$ were $E(20°) = 60.5 \times 10^6$ psi (418 GN/m²).

It should be noted that Equation (6) describes the total deformation strain which includes both the elastic and anelastic strain components. One can calculate the anelastic creep strain ε_a simply by replacing A_σ by $(A_\sigma - 1)$ in Equation (6). However in any practical fiber creep test, it would be very difficult to observe ε_a directly since like the elastic strain it develops linearly with stress and also begins to recover immediately upon stress removal. For this reason, total deformation strain is considered to be a more practical parameter for understanding and describing boron fiber creep. One final design point is that although considerations of stress effects on A_σ may be important in some situations, one could in many circumstances neglect the stress effects and design for the upper limit boron fiber creep by simply employing A_H in Equation (6).

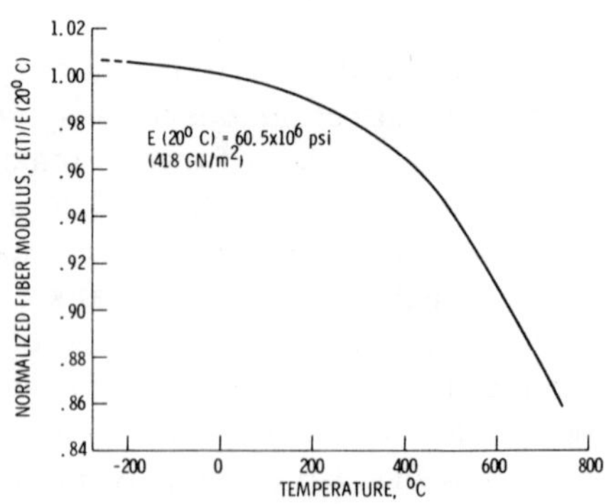

***Figure 3.** The normalized axial Young's modulus of boron fibers.*

Fiber fracture — Smith has observed that the two primary flaw sites responsible for crack propagation in commercially-produced boron fibers are located within the tungsten boride core and on the fiber surface [5]. By slightly etching the as-produced fibers, he was able to remove the surface flaws and thus observe only core flaw-initiated fiber fracture. As described earlier, one can utilize this fact to develop a "composite fiber" fracture model for calculating the high stress A_H from fracture stress data on slightly etched as-produced fibers. The basic equation for this model is Equation (4) in which it is assumed fiber fracture occurs at the core fracture strain which is time and temperature independent. Since fracture stress data were employed to determine A_H, it follows that Equation (4) can be transposed to predict $\bar{\sigma}_u$ as a fraction of time and temperature. That is, the average fracture stress of an etched fiber (core controlled fracture) can be calculated from the equation

$$\bar{\sigma}_u(q) = \bar{\sigma}_u(q_o) \frac{E(T)}{E(T_o)} \frac{A(q_o)}{A_H(q)} \tag{7}$$

Here $\bar{\sigma}_u(q_o)$ is the average fracture stress measured at some reference q_o condition, such as, a short-time tensile test ($t_o \simeq 60$ sec) at room temperature (T_o = 293 *K*). Although $A_H(q)$ was determined from short-time fracture test data, it should be realized that theoretically Equation (7) can be generalized through the q parameter to predict core-initiated fracture during other time-dependent tests such as impact and long-time stress rupture. At the present time, however, no data exist to confirm the validity of Equation (7) for other than the short-time tensile test.

The derivation of Equation (7) was simplified by the fact that the etched fiber could be treated as a two component composite in which the outer sheath component fractures whenever the inner core component reaches its fracture strain. This composite model may not be valid, however, when the source of fiber fracture are flaws on the fiber surface. In most as-produced commercial boron fibers, Smith [5] has observed only surface flaw and core-initiated fractures. He found that the two flaw types can be practically distinguished by the fact that core-initiated fractures generally produce strength values greater than 600 ksi (4.1 GN/m^2) whereas surface flaw-initiated fractures produce strength values less than 500 ksi (3.4 GN/m^2). Since commercial boron fiber spools are generally quoted at average strengths of 500 ksi, it follows that surface flaws do exist in these fibers. Thus, the question arises whether Equation (7) can be utilized to predict fracture stress of unetched as-produced fibers.

This question was examined empirically by plotting in Figure 4 the short-time temperature-dependent fracture stress data of Veltri and Galasso [9] for unetched as-produced boron fibers. To better compare these data with the theoretical predictions of Equation (7), the fracture stress values were normalized by dividing $\bar{\sigma}_u$ by the room temperature value, $\bar{\sigma}_u(q_o)$ = 500 ksi (3.4 GN/m^2). The range of the theoretical estimates based on Equation (7) and the errors in A_H (cf. Figure 1) are shown by the dashed lines. Comparing these with the experimental data, one finds that although surface flaws were most probably controlling fiber fracture, it does appear that Equation (7) predicts quite well the fracture stress of both etched and unetched as-produced fibers. From this result it follows that Equation (7) can be employed as the general equation for predicting boron fiber fracture stress regardless of the flaw type responsible for fracture initiation. However, it should be realized that anelastic creep effects on fiber fracture are contained in the A_H function used in Equation (7). Thus, in light of the stress effect curves of Figure 2, one should replace A_H by A_L if fiber flaws should initiate fracture at stress levels below the transition stress σ^*.

B/Al Composites

Fiber creep — The anelastic A function for boron fibers within B/6061 Al composites was determined from composite damping data [6]. The results are shown in Figure 5 as curve A_L^{II}. The subscript L again refers to the fact that low fiber stresses (<1 ksi (7 MN/m²)) were used for these measurements. The superscript II is used to distinguish the results for fibers within B/Al composites from the as-produced fiber results which are now labeled with the superscript I.

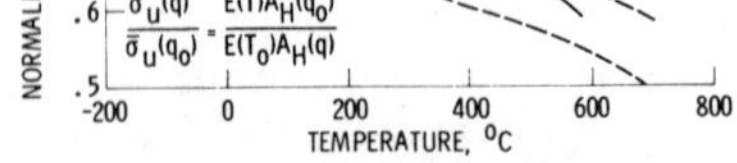

Figure 4. The normalized ultimate tensile strength of as-produced boron fibers.

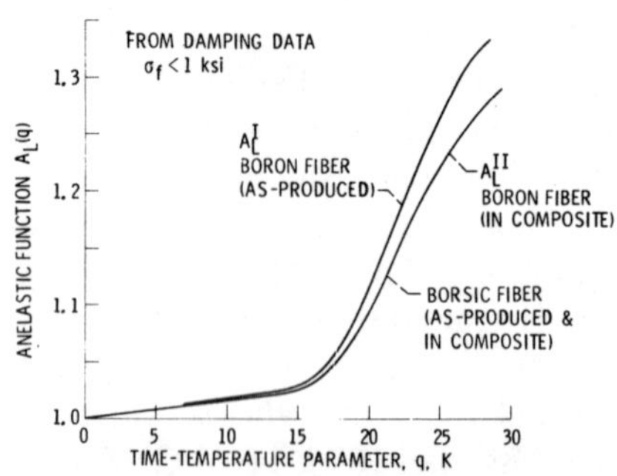

Figure 5. The effects of composite fabrication and SiC coating on the low-stress anelastic strain function for as-produced boron fibers.

Comparing the A_L^{II} curve with the as-produced A_L^I curve, which is also shown in Figure 5, one observes that the anelastic creep of the as-produced boron fibers is measurably reduced within B/Al composites. Obviously a microstructural change must have occurred in the fiber at some time during composite fabrication. Since the most adverse environmental conditions existed during high temperature diffusion bonding of the composite specimen, the microstructural change most probably occurred during this stage when the boron fibers reacted with the aluminum matrix to form the interfacial bond required for mechanical load transfer. In support of this surface reaction effect on the A function, it was found [7] that Borsic fibers in both the as-produced and composite conditions possess low stress A functions equivalent to the A_L^{II} result in Figure 5. Thus, on the basic level, the boron fiber microstructure responsible for its bulk anelastic deformation character can be measurably affected by surface reactions either with matrices or with fiber coatings. In terms of the boron unit sliding model, it would appear that at the high temperatures of the surface reactions, diffusional processes occur within the fiber which increase the number of immobilized boron units.

Although A_L^{II} can now be inserted into Equation (6) to accurately predict low-stress boron and Borsic creep in B/Al composites, the upper stress limit σ^* at which it can no longer be validly employed remains undetermined. As with the as-produced single fibers, high stress deformation experiments are required from which A^{II} versus stress data can be determined. These experiments could be performed either on single fibers extracted from B/Al composites or on unidirectional B/Al composites in which the fiber contributions can be easily and accurately measured. One obvious method of maximizing fiber effects and minimizing matrix effects in composites would be to study specimens with fiber volume fractions of 50 percent or greater. It should be mentioned that high stress creep data for 50 fiber volume percent B/Al composites do exist in the literature [10]. Attempts to employ these data for extracting high stress A^{II} curves were not very fruitful, primarily due to the existence of unknown matrix stress relaxation effects on total composite creep [11]. However, although composite creep data were not useful, it was found that a theoretical study of composite fracture stress versus temperature data could not only shed light on high stress effects on fiber creep but also yield equations for predicting time-temperature effects on B/Al fracture. The theory and results of this study will now be discussed.

Composite fracture — Having established that the A_H function is the appropriate function for predicting high stress fracture of single as-produced fibers, the first question to be answered is whether this function can be used also to predict the high stress creep and fracture of boron fibers within B/Al composites. To examine this question, a literature search was conducted for short-time fracture stress versus temperature data for B/6061 Al composites. In order to minimize matrix loading effects, the search was confined only to composites with nominally 50 percent fiber volume fraction. Summary plots of the literature data which fit this requirement are shown in Figure 6. Although the tensile strengths vary in magnitude from one source to another, one can notice definite trends in the temperature-dependent behavior. For example, below 200 °C the data from all sources indicate essentially no dependence on temperature. Above 200 °C the strength data fall off reaching about 80 percent of the room temperature value near 300 °C.

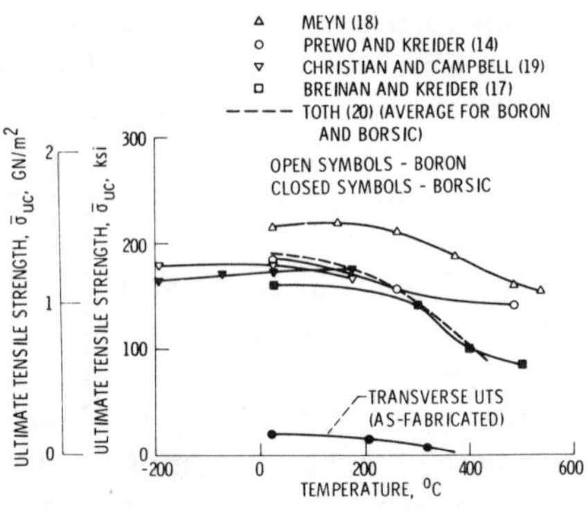

Figure 6. The ultimate tensile strength of B/6061 Al composites.

Regarding fracture of fibers within B/Al composites, one might as a first approximation neglect matrix contributions and assume that the composite stress should drop off at least as rapidly as the fracture stress for single as-produced boron fibers. Examining the experimental results of Figure 4, it is seen that single as-produced fibers fall off to 80 percent of $\bar{\sigma}_u(q)$ near 150° C, a much faster dropoff than the composite. Thus, from a practical point of view, it appears that the Veltri and Galasso data for as-produced fibers cannot be used to understand and predict the temperature dependence of composite strength. From an analytical point of view, one must conclude that the insertion of the A_H in the fiber fracture theory of Equation (7) will not explain the Figure 6 composite data. Since the A_L results of Figure 5 indicate that boron fibers will creep within B/Al composites, the problem of predicting the effects of time and temperature on B/Al axial fracture thus becomes one of not only accounting for matrix plasticity but also of determining the appropriate A function for high stress creep and fracture of the fibers after composite fabrication.

To solve this problem, a theoretical analysis was made of the major factors which affect the temperature-dependent behavior of the fracture stress of metal matrix composites in general and B/6061 Al composites in particular. The axial fracture model chosen is that developed by Rosen [12] in which the fracture modes of fibers within an unidirectional composite are controlled by the fracture of fiber bundles. That is, the composite or fiber bundle completely fails when enough fiber breaks occur so that the load carried by the remaining intact fibers exceeds their strength capability. Due to a distribution of fiber strengths, the weak fibers fracture first leaving the stronger fibers to carry the load. Common practice is to describe the distribution in fiber strengths according to a Weibull distribution [13]. In this case, the average fiber strength σ_{uf} and the average fiber bundle strength σ_{bf} are given by

$$\bar{\sigma}_{uf} = \sigma_o \left[\frac{L_s}{d}\right]^{-1/\omega} \Gamma\left(\frac{\omega + 1}{\omega}\right) \tag{8}$$

and

$$\bar{\sigma}_{bf} = \sigma_o \left[\frac{e\omega L_b}{d}\right]^{-1/\omega} \tag{9}$$

Here σ_o is the Weibull distribution scale factor, ω is the Weibull shape factor which describes the scatter in strength values, Γ is the tabulated gamma functin, e is the natural base constant, d is the fiber diameter, and L_s and L_b are the test gage lengths of the single fiber and the fiber bundle, respectively. Typically $\omega > 1$, so that $\bar{\sigma}_{bf}$ decreases as the bundle length increases.

When a fiber breaks in a metal matrix composite, the matrix by virtue of its plastic character localizes the loss of load carrying ability of the broken fiber. That is, at an axial distance $\delta/2$ on either side of the break, the stress in the broken fiber returns to the average stress of all the intact fibers. Because of the existence of this "ineffective" length δ, Rosen considers the composite to be made of a series of independent fiber bundles each of length δ. As such, total fracture of the composite occurs whenever any one of these short length bundles fail. If one assumes perfectly plastic behavior for the matrix (no workhardening effects), the length δ of the bundles can be calculated from

$$\delta = \sigma d/2\tau_m \tag{10}$$

where $\sigma = \bar{\sigma}_{bf}$ is the stress in the fibers at composite fracture and τ_m is the shear strength of the matrix. Since τ_m decreases with temperature, it follows that δ will increase, giving rise by Equation (9) to a reduction in $\bar{\sigma}_{bf}$ with temperature.

To express the total stress level at which a metal matrix composite will fracture, one can employ the above concepts to write the following rule-of-mixtures equation for the average (ultimate) tensile strength of the composite

$$\bar{\sigma}_{uc} = v_f \bar{\sigma}_{bf}(\delta) + v_m \sigma_{ym}(T) \tag{11}$$

Here v_f and v_m are the volume fraction of the fiber and matrix, respectively, and σ_{ym} is the tensile yield strength of the matrix. Equation (11) neglects stress concentration effects of broken fibers on nearby fibers. It also neglects residual stress effects on the fibers due to cooling the composite from fabrication temperature. For soft matrices such as aluminum these effects are small and also tend to oppose each other as temperature is varied. For calculating the temperature dependence of $\bar{\sigma}_{uc}$ it is convenient to normalize each stress term in Equation (11) by dividing by its value for a short-time tensile test at room temperature. The following R parameters are thus defined:

$$R_{uc} = \bar{\sigma}_{uc}(q)/\bar{\sigma}_{uc}(q_o) \tag{12}$$

$$R_{bf} = \bar{\sigma}_{bf}(q)/\bar{\sigma}_{bf}(q_o) \tag{13}$$

$$R_{ym} = \sigma_{ym}(T)/\sigma_{ym}(T_o) \tag{14}$$

In anticipation of applying these equations to B/A1 composite behavior, the temperature variable T was replaced, where appropriate, by the more general q variable. Reference conditions (q_o and T_o) are taken as a short-time ($t_o \simeq 60$ sec) tensile test at room temperature ($T_o \simeq 293\,°K$). Inserting Equations (12), (13), and (14) into Equation (11) one obtains

$$R_{uc} = (R_{bf} + \beta R_{ym})/(1 + \beta) \tag{15}$$

where the constant $\beta = v_m \sigma_{ym}(T_o)/v_f \bar{\sigma}_{bf}(q_o)$. Thus, to predict composite fracture stress $\bar{\sigma}_{uc}(q)$, one simply needs a theoretical formula for R_{bf} plus experimental information for $\bar{\sigma}_{uc}(q_o)$ and $R_{ym}(T)$.

To derive R_{bf} one can utilize Equation (9) to write

$$R_{bf} = \frac{\sigma_o(q)}{\sigma_o(q_o)} \left[\frac{\bar{\delta}(T)}{\delta(T_o)}\right]^{-1/\omega} \tag{16}$$

Here it is assumed that the flaw distribution as measured by ω does not change during the test. Also, to include fiber creep effects, a q dependence has been assigned to σ_o since by Equation (8), σ_o is the only term responsible for a change in fiber strength with temperature. Assuming matrix shear strength τ_m is directly proportional to matrix yield strength σ_{ym}, one finds from Equation (10) that

$$\delta(T)/\delta(T_o) = R_{bf}/R_{ym} \tag{17}$$

By inserting this result into Equation (16) and manipulating, it follows that

$$R_{bf} = \left[\frac{\sigma_o(q)}{\sigma_o q_o)}\right]^{\omega/(1+\omega)} \left[R_{ym}\right]^{1/(1+\omega)} \tag{18}$$

Thus, under the assumptions stated above, one should now be able to employ Equations (15) and (18) to predict the time-temperature dependent fracture stress of metal matrix composites. For those composites reinforced by boron fibers, both terms on the right hand side of Equation (18) must be considered. However, for those composites reinforced by elastic fibers, $\sigma_o(q) = \sigma_o(q_o)$ so that only the second matrix-related term need be considered.

To determine the high stress A function for boron fibers in B/Al composites, Equations (15) and (18) were applied to the experimental strength data of Figure 6. It was assumed that the temperature dependence of R_{ym} for the 6061 aluminum matrix followed that measured by Prewo and Kreider [14] for the transverse tensile strength of as-fabricated B/6061 composites. Their results which are plotted in Figure 6 show a drop in transverse strength from 20 ksi (138 MN/m²) at room temperature to an extrapolated value of 3 ksi (21 MN/m²) at 370° C [21]. Assuming a room temperature matrix shear strength of 14 ksi (97 MN/m²) [15] and a 5.6 mil (142 μm) diameter fiber whose strength falls off by 70 percent at 370 °C, it follows from Equation (10) that the ineffective length δ increases from 0.1 to 0.5 inch (2.5 to 12 mm) between 20° and 370° C. To compare the Figure 6 strength data with theory each set of data was normalized by dividing by their room temperature values and the best fit R_{uc} for all sets was calculated. The best fit result is shown by the dashed curve of Figure 7.

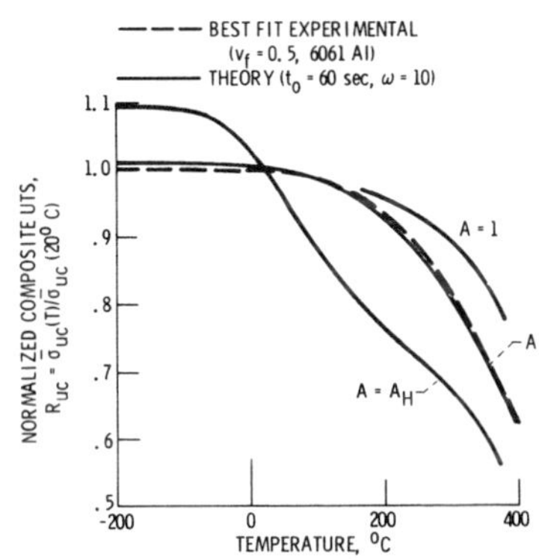

Figure 7. The normalized ultimate tensile strength of B/6061 A1 composites.

Regarding values for the parameters in the theoretical equations, σ_{bf} and σ_{ym} at room temperature were assumed to be 400 and 8 ksi (2.8 and 0.06 GN/m²) [14], so that $\beta = 0.02$ for 50 percent fiber content in as-fabricated 6061 A1. In examining the literature for information concerning the Weibull distribution of boron fibers removed from B/6061 A1 composites, ω parameters were found which range between 8 and 12 with $\omega = 10$ as an average value [14,15,16]. Since the boron fibers definitely show anelastic creep within the composites, the assumption was made that fiber fracture could be described by Equation (7) with an appropriate A function. Thus, for Equation (18), one can write

$$R_{bf} = \left[\frac{E(T)\,A(q_o)}{E(T_o\,A(q)}\right]^{\omega/(1+\omega)} \left[R_{ym}\right]^{1/(1+\omega)} \tag{19}$$

With $\omega = 10$, R_{bf} was inserted into Equation (15) and R_{uc} was calculated for three different A functions: $A = 1$ which assumes strictly elastic fiber behavior (no creep), $A = A_H$ which assumes the as-produced high stress condition for the fibers, and $A = A_L^{II}$ which assumes that the low stress creep behavior does not change for fiber stresses up to ~400 ksi (2.8 GN/m²).

The theoretical results for the three A functions are shown in Figure 7. For elastic fiber behavior, curve $A = 1$ clearly shows that composite strength should fall off simply due to matrix effects in which temperature decreases τ_m and increases the ineffective length δ. If $\omega = \infty$, which implies constant strength fibers, Equation (19) indicates that the matrix would have no effect on fiber bundle strength. For the assumption of as-produced behavior, the A_H curve obviously drops off much too rapidly to explain the as-produced fiber data of Figure 4 with the composite data of Figure 6. Finally, it appears that the A_L^{II} curve gives the best fit to the fracture data. Thus it may be reasonable to assume that fiber creep within B/A1 composites is governed by only the A_L^{II} function from zero to at least 400 ksi (2.8 GN/m²). The shift from A_L to A_H observed for the as-produced fiber (cf. Figure 2) apparently does not occur for the B/A1 fibers. From a basic point of view one might interpret this as an additional effect of the boron-aluminum interfacial reaction in that besides decreasing the number of mobile boron units, it may also affect the internal locking mechanism in such a way as to increase σ^* to values greater than 400 ksi.

Summarizing the design aspects of the above discussion, it appears that for any fiber stress less than 400 ksi (2.8 GN/m²), the A_L^{II} curve of Figure 2 is the proper A function for describing boron and Borsic creep in B/6061 A1 composites. Thus, Equation (6) with $A_\sigma = A_L^{II}$ is the predictive equation for fiber creep. Regarding composite fracture, Equations (15) and (18) with $A = A_L^{II}$ and $\omega = 10$ give good estimates of B/6061 A1 axial fracture strength as a function of temperature. However, because of the thermally-activated nature of boron creep, these equations can be generalized through the q parameter to also include test conditions in which time is the principle variable. For example, in an axial stress rupture test, one can assume $\beta R_{ym} = 0$ (due to rapid stress relaxation in the matrix) and write for the stress rupture strength

$$\bar{\sigma}_{uc}(q) = \bar{\sigma}_{uc}(q_1) \left[\frac{A(q_1)}{A(q)}\right]^{\omega/(1+\omega)} \tag{20}$$

Here $q \geqslant q_1$ and $\bar{\sigma}_{uc}(q_1)$ is the short-time composite strength at temperature T_1 which is held constant during the test. Since q develops with the log of time, $\bar{\sigma}_{uc}$ should show only a weak time dependence. Indeed, limited stress rupture data for B/A1 composites confirm such a dependence [17].

The apparent confirmation of the mechanistic theory used to calculate composite fracture stress suggests possible methods of improving the creep and associated stress rupture characteristics of B/Al composites. For example, the fact that fiber creep is reduced by high temperature surface reactions with an SiC coating and a 6061 aluminum matrix suggests that perhaps other fiber secondary treatment processes exist which can either eliminate of drastically reduce anelastic creep in the as-produced boron fibers. These treatment processes, however, should have minimum adverse effects on the as-produced fiber flaw character or else the benefits gained by reducing creep could be lost by a degradation in fiber strength. Regarding the matrix shear strength effect on composite fracture, it follows from Equation (18) that this effect can be minimized by narrowing flaws within the interfacial surface phase that control fiber fracture in metal matrix composites, the largest ω values should result from those bonding conditions which produce the most uniform distribution of interfacial flaws. Thus, a good mechanical interfacial bond may not only produce a stronger composite at room temperature but also at higher temperatures where matrix shear strength falls off.

SUMMARY OF RESULTS AND CONCLUSIONS

Due to linear stress dependence of anelastic strain, it is possible to write simple analytical equations for describing boron fiber creep and fracture as a function of time, temperature, and stress. The primary time-temperature dependent variable in these equations is an anelastic strain function A. From analysis of single fiber deformation and fracture data, A functions for commercial boron fibers were determined. Thus, the creep strain and stress rupture strength of as-produced boron fibers can now be predicted.

Analysis of Borsic fiber and B/Al composite damping data indicates that the creep of SiC coated boron fibers and of boron fibers in B/Al composites is measurably less than the creep of as-produced boron fibers. The reduced A function characteristic of fiber creep in B/Al composites was determined. This function together with a general metal matrix composite fracture theory should now allow fairly good estimates of the effects of time and temperature on the axial failure modes of B/Al composites. Based on the mechanistic models employed in this composite fracture theory, it appears that a good interfacial fiber-matrix bond will not only maximize mechanical load transfer but also may reduce detrimental fiber and matrix effects on B/Al stress rupture properties.

NOMENCLATURE

$A{:}A_L,A_H$	Anelastic strain function: at low stress, at high stress
d	fiber diameter
E	fiber Young's modulus
L_s,L_b	gage length of single fibers and fiber bundle
q	time-temperature parameter

R_{bf}, R_{uc}, R_{ym}	ratio of test value to short-time value at room temperature for fiber bundle strength, composite tensile strength, and matrix yield strength
t	deformation time (sec)
T	test temperature (K)
v_f, v_m	volume fraction of fibers and matrix
β	ratio of matrix load to fiber load under short-time composite loading at room temperature
δ	"ineffective length" of fiber bundles in composites
$\varepsilon{:}\varepsilon_e,\varepsilon_a$	total fiber strain: elastic component, anelastic component
$\bar{\varepsilon}_u$	average (ultimate) fiber fracture strain
σ_o	Weibull distribution stress factor
σ^*	transition stress for change from A_L to A_H
$\bar{\sigma}_{uf}$	average (ultimate) fiber tensile strength
$\bar{\sigma}_{bf}$	average fiber bundle strength
$\bar{\sigma}_{uc}$	average (ultimate) composite tensile strength
σ_{ym}	tensile yield strength for matrix
τ_m	shear strength of matrix
ω	Weibull distribution shape factor

Superscripts

$I;II$	for as-produced fibers; for fibers within B/A1 composites

REFERENCES

1. J.A. DiCarlo: "Time-Temperature-Stress Dependence of Boron Fiber Deformation," *Composite Materials: Testing and Design (Fourth Conference)*, AASTM STP 617, pp. 443-465, American Society for Testing Materials, Philadelphia, 1977.
2. K.M. Prewo: "Anelastic Creep of Boron Fibers," *J. Compos. Mater.*, 1974, vol. 8, pp. 411-414.
3. J.A. DiCarlo: "Anelastic Deformation of Boron Fibers," *Scripta Met.*, 1976, vol. 10, pp. 115-119.
4. J.A. DiCarlo: "Mechanical and Physical Properties of Modern Boron Fibers," *ICCM/2, Second International Conference on Composite Materials,* pp. 520-538, The Metallurgical Society of AIME, New York, 1978.
5. R.J. Smith: "Changes in Boron Fiber Strength due to Surface Removal by NASA TN D-8219, Washington, D.C., 1976.
6. J.A. DiCarlo and J.E. Maisel: "Measurement of the Time-Temperature Dependent Dynamic Mechanical Properties of Boron/Aluminum Composites," *Composite Materials: Testing and Design (Fifth Conference)*, ASTM STP 674, pp. 201-227, American Society for Testing Materials, Philadelphia, 1979.
7. J.A. DiCarlo and W. Williams: "Dynamic Modulus and Damping of Boron Silicon Carbide, and Alumina Fibers," NASA TM-81422 Washington, D.C., 1980.
8. T.E. Firle: "Amplitude Dependence of Internal Friction and Shear Modulus of Boron Fibers," *J. Appl. Phys.*, 1968, vol. 39, pp. 2839-2845.
9. R.D. Veltri and F.S. Galasso: "High-Temperature Strength of Boron, Silicon Carbide, Stainless Steel, and Tungsten Fibers," *J. Am. Ceram. Soc.*, 1971, vol. 54, pp. 319-320.

10. K.G. Kreider and K.M. Prewo: "Boron-Reinforced Aluminum," *Composite Materials,* K.G. Kreider, ed., Vol. 4, pp. 399-471, Academic Press, New York, 1974.
11. A.R.T. de Silva: "A Theoretical Analysis of Creep in Fibre Reinforced Composites," *J. Mech. Phys. Solids,* 1968, vol. 16, pp. 169-186.
12. B.W. Rosen: "Tensile Failure of Fibrous Composites," *AIAA J.,* 1964, vol. 2, pp. 1985-1991.
13. H.T. Corten: "Micromechanics and Fracture Behavior of Composites," *Modern Composite Materials,* L.V. Broutman and R.H. Krock, eds., pp. 27-105, Addison-Wesley, Reading, Mass., 1967.
14. K.M. Prewo and K.G. Kreider: "High Strength Boron and Borsic Fiber Reinforced Aluminum Composites," *J. Compos. Mater.,* 1972, vol. 6, pp. 338-357.
15. M.A. Wright and J.L. Wills: "The Tensile Failure Modes of Metal Matrix Composite Materials," *J. Mech. Phys. Solids,* 1974, vol. 22, pp. 161-175.
16. M.A. Wright and D. Welch: "Failure of Centre Notched Specimens of 6061 Aluminum Reinforced with Unidirectional Boron Fibers," Fibre Sci. Technol., 1978, vol. 11, pp. 447-461.
17. E.M. Breinan and K.G. Kreider: "Axial Creep and Stress-Rupture of B-Al Composites," *Metall. Trans.,* 1970, vol. 1, pp. 93-104.
18. D.A. Meyn: "Effect of Temperature and Strain Rate on the Tensile Properties of Boron-Aluminum and Boron-Epoxy Composites," *Composite Materials: Testing and Design (Third Conference),* ASTM STP 546, pp. 225-236, American Society for Testing Materials, Philadelphia, 1974.
19. J.L. Christian and M.D. Campbell: "Mechanical and Physical Properties of Several Advanced Metal-Matrix Composite Materials," *Cryogenic Engineering Conference, Advances in Cryogenic Engineering,* Vol. 18, pp. 175-183, Plenum Press, New York, 1973.
20. I.J. Toth: *Composite Materials:* "Comparison of the Mechanical Behavior of Filamentary Reinforced Aluminum and Titanian Alloys," *Testing and Design (Third Conference),* ASTM STP 546, pp. 542-560, American Society for Testing Materials, Philadelphia, 1974.
21. *Metals Handbook,* Vol. 1, p. 946, American Society of Metals, Metals Park, Ohio, 1961.

On the Specimen Dependence of Unidirectional Boron/Aluminum Fracture Toughness*

E. D. REEDY, JR.

*Sandia Laboratories***
Albuquerque, New Mexico 87185

ABSTRACT

Fracture toughness, as commonly defined for unidirectional boron/aluminum, was measured with center-notched, compact tension, and various sized three-point bend specimens. All coupons were cut from the same plate of material. The values of fracture toughness determined were found to be specimen dependent, thus indicating that fracture toughness is not a material property for the size and type of specimens tested. The failure mode was also found to be specimen dependent. Crack growth colinear with the notch occurred in the center-notched and three-point bend specimens, while the crack initiated and propagated perpendicular to the notch tip in compact tension specimens. The failure of linear elastic fracture mechanics to define a material property for these specimens was attributed to extensive matrix yielding in regions remote from the notch tip. The existence of widespread matrix yielding was confirmed with a strain gaged, center-notched tensile specimen.

INTRODUCTION

THERE HAVE BEEN frequent attempts to apply the concepts of linear elastic fracture mechanics to transversely notched, unidirectional boron/aluminum [1–17]. A plane strain value for fracture toughness, as defined by a strict interpretation of the ASTM standard test method E399-78 [18], does not exist for this material. The standard test method requires that an adequately sized notched specimen be pre-cracked in fatigue, in a specified manner, to introduce a sharp crack colinear with the original notch. This cannot be done with transversely notched, unidirectional boron/aluminum. When this composite is fatigued, the notch is blunted by shear-induced matrix damage parallel to the fibers [3, 19]. The standard test method also places requirements on the allowable form of the load versus crack opening displacement record to insure crack tip yielding is localized. Published

*This work was supported by the U.S. Department of Energy (DOE), under Contract DE-AC04-76-DP00789.

**A U.S. DOE facility.

boron/aluminum load versus crack opening displacement records do not in general satisfy those requirements [1, 4, 7, 8, 12, 20].

Experimenters studying transversely notched, unidirectional boron/aluminum often modify the standard test method for determining plane strain fracture toughness to allow for the composite's nature. Notched specimens are tested without fatigue precracking. Frequently electrical discharge machining is used to introduce notches with a root radius on the order of 0.1 mm. The load versus crack opening displacement record is used to identify one of two loads (Fig. 1). The load here identified as P_J corresponds to an initial discontinuity in slope. This load is not necessarily associated with visible notch growth, but discontinuities in the slope are presumed to be a consequence of fiber breakage. This value is not always well defined. The maximum load level attained, P_M, is easily defined. The formula relating load, crack length, and specimen geometry to the crack tip stress intensity factor for a linear elastic, isotropic material [21] is used to calculate the fracture toughness. The load level P_J or P_M is used in this relationship to define a critical fracture toughness K_J or K_M respectively. Either the initial notch length or an effective notch length which corrects for notch tip yielding is employed in the calculation.

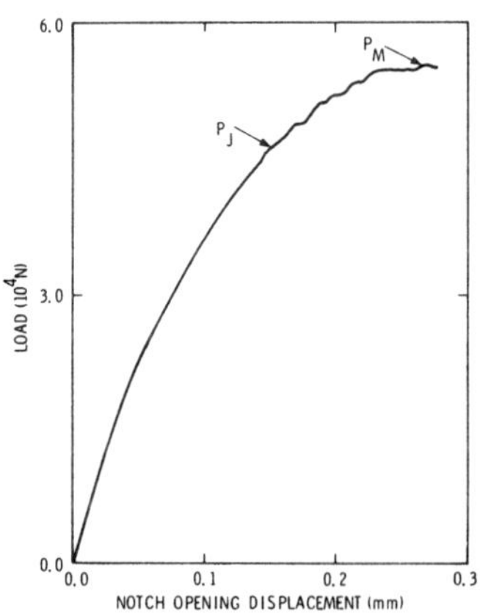

***Figure** 1. Typical load vs. notch opening displacement record for a center-notched specimen with a/w = 0.5.*

The fracture toughness (notched strength) of boron/aluminum, as defined above, has been found to be independent of notch root radius [4, 7, 22] and also of plate thickness when it exceeds 2. mm [3, 4, 7, 9, 12]. These observations suggest that it may not be necessary to introduce a sharp fatigue crack colinear to the notch and that relatively thin plates of boron/aluminum can be used to determine a plane strain fracture toughness. It has also been observed that, for a fixed specimen geometry, the fracture toughness is constant over a wide range of intermediate notch lengths [7, 8, 9, 10, 12, 14]. However, fracture toughness K_J or K_M has yet to be established as a material property.

Published fracture toughness values for 50% volume fraction 142 μm diameter boron/6061 aluminum in the as fabricated condition are displayed in Table 1. These data were determined from specimens with a thickness exceeding 1.9 mm. Ianuzzi [5] and Hoover [8] determined a fracture toughness based upon a "pop-in" or slope discontinuity in the load vs. notch opening displacement record. The value K_J determined with a center-notched specimen (203.2 mm long by 25.4 mm wide) exceeds by 40% that measured with a small three-point bend specimen (63.5 mm long by 10.2 mm high

Table 1. Literature Values for Toughness of Transversely Notched Unidirectional 142 μm B/Al-6061-F with $V_F = 0.5$ and $T \geq 1.9$ mm

Specimen (L x W x T / mm x mm x mm)	K_J MPa $\sqrt{m}$	K_M MPa $\sqrt{m}$	Reference
CN (203.2 x 25.4 x 2.2)	54.	75.	Iannuzzi [5]
TPB (63.5 x 10.2 x 5.1)	39.	---	Hoover [8]
CT (61.0 x 101.6 x 3.8, 7.6)	cracked parallel to fibers		Sun and Prewo [11]
TPB (127.0 x 25.4 x 1.9, 3.8, 7.6)	---	76.	Waszczak [12]
CN (254.0 x 25.4 x 1.9)	---	83.	
CN (76.2* x 25.4 x 25.4)	---	73	Wright and Welch [15,16,17]
CN (152.4* x 50.8 x 25.4)	---	73.	
CN (304.8* x 101.6 x 25.4)	---	58.	

CN - Center-notched
TPB - Three-point bend
CT - Compact tension
*Gage length

with 39.9 mm span). Most of the fracture toughness values reported are based upon the maximum load attained during the test. Iannuzzi [5], Waszczak [12], and Wright and Welch [15, 16, 17] tested center-notched specimens to determine K_M. The values determined for specimens with a 25.4 mm width ranged from 73 to 83 MPa $\sqrt{m}$. Data recently published by Wright and Welch [15, 16, 17] suggest that K_M may decrease by 20% when the absolute dimensions of center-notched specimens, with a gage length-to-width ratio equal to 3, are increased by a factor of four. Wasczak [12] tested two different specimen types and found consistent values of K_M for 25.4 mm wide center-notched specimens and large three-point bend specimens (127. mm long by 25.4 mm high with a 101.6 mm span). Notch growth in center-notched and three-point bend specimens was always roughly colinear with the original notch. However, the compact tension specimens tested by Sun and Prewo [11] exhibited crack growth parallel to the fibers which is perpendicular to the original notch.

The above results suggest that fracture toughness as commonly defined for boron/aluminum may be dependent on specimen type and dimensions. However, since the tests were conducted by 5 different researchers over a period exceeding 5 years, no firm conclusion can be stated. The materials used were of different thicknesses, secured from various suppliers, as well as prepared and tested with different procedures. This study attempts to resolve that uncertainty. Specimens with dimensions similar to those listed in Table 1 were fabricated from a single plate of 142 μm boron/6061 aluminum and tested. Sufficient numbers of each specimen geometry were tested to insure a statistically significant result.

EXPERIMENTAL PROCEDURES

Material

The unidirectional composite system investigated had a nominal 50% volume fraction of 142 μm boron fibers in a 6061 aluminum matrix. The material was supplied by DWA, Chatsworth, California, and tested in the as-fabricated condition. All specimens were cut from a single plate with a 610 mm length in the fiber direction, 305 mm width, and approximate 5.1 mm thickness. The material was inspected to insure its acceptability. Micrographs showed the material to be fully bonded and a chemical analysis indicated a nominal 50% volume fraction of fibers. The composite strength in the fiber direction as measured by a 4-point bend test was 1690 MPa while its transverse tensile strength was 130 MPa. Ten center-notched, 3 compact tension, together with 8 small, 13 medium and 9-large three-point bend specimens were cut from the plate with a diamond impregnated cut-off wheel. The specimen dimensions are given in Figure 2. Notches and holes were introduced by electrical discharging machining techniques. The dimensions, notch length, and notch root radius of each specimen was measured prior to testing. The notch root radius varied from 0.1 mm to 0.15 mm.

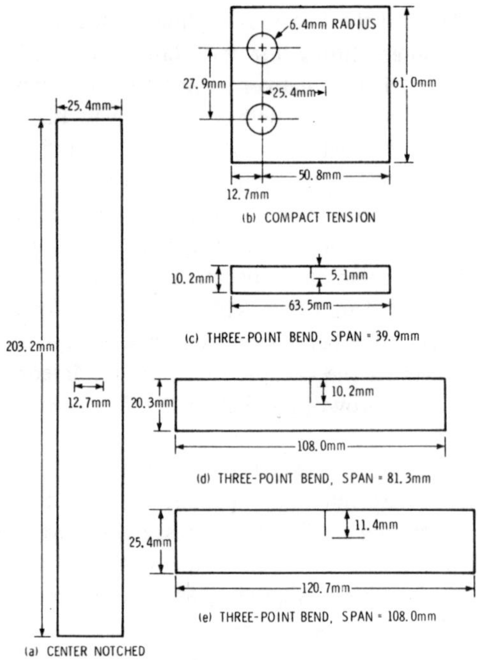

Figure 2. Fracture toughness specimens, all 5.1 mm thick.

Mechanical Testing and Data Reduction

The center-notched specimens were tested in tension using wedge action grips. The severity of the notch permitted the tests to be conducted without doubler tabs. A conventional clip gage was held by knife edges bonded to the specimen. Load versus notch opening displacement was recorded. The compact tension specimens were loaded through a clevis and pin arrangement. Knife edge supports for a clip gage were attached symmetrically about the notch on the specimen's edge. Load versus notch opening displacement was also recorded for this specimen. The majority of the three-point bend specimens were tested in a manner similar to that used by Hoover [8]. A three-point bend fixture with a span defined by fixed supports was used. All contact points were curved surfaces with at least a 3.2 mm radius. The midspan deflection was measured with a linear variable differential transformer located beneath the specimen and directly opposite the loading ram. Load versus midspan deflection was recorded on a *x-y* recorder. Several three-point bend specimens were also tested on a bend fixture with roller supports as suggested by the ASTM plane strain fracture toughness standard test method [18]. A clip gage was used with these specimens to record load versus notch opening displacement.

The data was analyzed in the manner described in the introduction of this report. The load P_J, corresponding to a slope discontinuity in the load-displacement trace, and also the maximum load attained, P_M, were identified. These load values, plus the previously measured notch length and specimen dimensions, were used to calculate the fracture toughness K_J or K_M. The relationship defining the stress intensity of an isotropic material as given by Tada [21] was used for the calculation. The suitability of applying a relationship determined for an isotropic material to an orthotropic material such as boron/aluminum was investigated. Orthotropic, linear elastic finite element calculations were performed for boron/aluminum center-notched, compact tension, and three-point bend specimens. These calculations employed a computer program developed by Benzley and Beisinger [23] which incorporates the correct form of the crack tip singularity in a linear elastic, orthotropic material. The calculated stress intensities were always within 3% of isotropic values.

The data were also analyzed in a manner similar to that employed by Awerbuch and Hahn [14], who utilized a method suggested by the work of Waddoups, Eisenmann, and Kaminski [24] and Nuismer and Whitney [25]. In essence, an effective notch length is substituted for the initial length in the calculation of the fracture toughness. This fracture toughness K_M^* is defined as

$$K_M^* = Y\sigma_M \sqrt{\pi (a + C_o)} \tag{1}$$

where

Y	– specimen finite width correction for isotropic stress intensity
σ_M	– notched strength
σ_o	– unnotched strength
a	– notch length/2 for symmetric specimens, otherwise notch length

and

$$C_o = a / \left[\left(\frac{\sigma_o}{Y\sigma_M} \right)^2 - 1 \right]$$

The quantity C_o is assumed to be a material property which could be determined from a single test. It may be considered to represent an inherent flaw size. In this paper an average value of C_o was determined from the tests of center-notched specimens; this value was then assumed valid for the other specimen types.

RESULTS AND DISCUSSION

The fracture toughnesses K_J, K_M, and K_M^* as determined by testing center-notched, various sized three-point bend, and compact tension specimens are recorded in Table 2. A value of C_o = 0.604 mm was determined from the averaged

Table 2. Toughness of Transversely Notched, Unidirectional 142 μm B/Al-6061-F with $V_F = 0.5$ and Crack Length/Width = 0.5

Specimen (L x W x T, mm x mm x mm)	Number Tested	K_J - MPa $\sqrt{m}$ (Mean with 95% Confidence Interval)	K_M - MPa $\sqrt{m}$ (Mean with 95% Confidence Interval)	K_M^* - MPa $\sqrt{m}$ (defined by (1))
CN (203.2 x 25.4 x 5.1)	10	54.3 ± 3.0	70.7 ± 3.4	76.8
TPB (63.5 x 10.2 x 5.1) Span = 39.9 mm	8	40.5 ± 0.6	42.2 ± 0.7	52.6
TPB (108.0 x 20.3 x 5.1) Span = 81.3 mm	8	48.8 ± 2.3	53.2 ± 1.5	58.6
TPB* (120.7 x 25.4 x 5.1) Span = 108.0 mm	6	64.5 ± 2.9	68.1 ± 1.3	72.9
CT (61.0 x 63.5 x 5.1)	3	---- Cracked Parallel to Fibers	32.9 ± 2.0	34.1

CN - center-notched
TPB - three-point bend
CT - Compact tension
*crack length/width = 0.45

center-notched tensile data and was used to calculate K_M^* (Eq. 1). The values of K_J, K_M, and K_M^* all depend on the type and size of specimen used in the measurement. The use on an effective notch length, as defined in Eq. 1, in place of the initial notch length in the calculation of fracture toughness did not substantially alter the specimen dependence of the toughness. All three methods of defining a fracture toughness show a similar variation with specimen type and size, consequently, only the fracture toughness K_M will be discussed.

The mean values of K_M show a marked specimen dependence. The 95% confidence intervals (constructed using Student's t-distribution; see *Introductory Probability and Statistical Applications,* Paul L. Meyer, Addison-Wesley, C 1970, p. 306) for these values are sufficiently tight to indicate that this variation is genuine. In fact, only two of the five specimens have overlapping scatter bands (Fig. 3). The center-notched specimens have an average value of K_M = 70.7 MPa $\sqrt{m}$. On the other hand the smallest three-point bend specimens (span = 39.9 mm) have an average value which is only 60% of the center-notched specimens. The fracture toughness of the three-point bend specimens increased with specimen size, the largest specimens having an average value K_M = 68.1 MPa $\sqrt{m}$. This value is slightly less than that of the center-notched specimens, and the 95% confidence intervals for their respective means overlap. Unlike the other specimen types, the compact tension specimens did not experience colinear notch growth. For these specimens failure occurred when a crack initiated and propagated at the notch tip in a direction perpendicular to the notch – in a direction parallel to the fibers. The formal use of the definition of fracture toughness K_M based on the maximum load attained yields an average value of 32.9 MPa $\sqrt{m}$ for this specimen. This is the lowest value measured for any of the specimens tested. The existence of two different failure modes, either colinear notch growth or perpendicular crack initiation and propagation, raises doubts to the existence of a single parameter characterizing failure. These tests show that the commonly used definitions for fracture toughness of unidirectional boron/aluminum do not define a material property which is independent of the type and size of specimens tested. It is also interesting that a comparison of Table 1 and 2 reveals that the fracture toughness values determined on specimens cut from a single plate agrees favorably with those values determined previously in separate studies.

As is seen in Figure 3, the scatter band for the center-notched specimens is much greater than that for the other specimen types. This is attributed in part to the difficulty of loading the specimen without introducing unintended stresses. Bending stresses can be introduced into the specimen when the center of the notch does not fall on the specimen's vertical center-line. Even if the specimen had complete geometric symmetry, notch growth probably will not be symmetric. Non-uniformity in grip pressure on the specimen may also cause variation in notch strength. It was observed from the imprint of the serrated grips on 4 of the 10 center-notched specimens that these specimens were held more firmly at their edges than at the

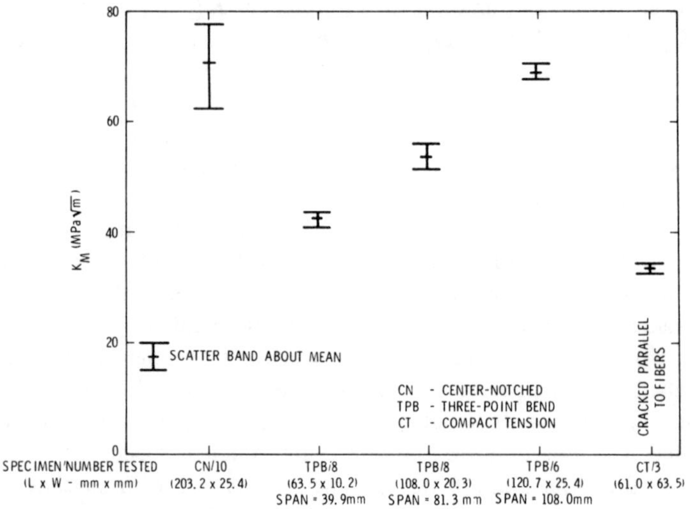

Figure 3. *Fracture toughness K_M of 142 μm B/Al-6061-F for various 5.1 mm thick specimens.*

center. The average value of K_M for these 4 specimens was approximately 10% below that based upon the other six specimens. If inadvertent bending stress could be eliminated, and if the load could be uniformly applied, then the mean value of K_M might be increased by a few percent – perhaps 5% to 10% – over that reported in Table 2, and data scatter would be reduced. In any event, the mean value of the center-notched specimen fracture toughness K_M exceeds that of any other specimen type tested.

The effect of three-point bend fixture design on the measured value of K_M was also examined. Initial tests used a fixture similar to that employed by Hoover [8] which utilized fixed supports. The loads required to fracture larger bend specimens caused noticeable plastic deformation (brinelling) under the loading ram and supports. This brinelling could indicate substantial lateral friction at the supports, which might increase the load required to propagate the notch. To investigate that possibility, 5 medium (108.0 mm × 20.3 mm × 5.1 mm) and 3 large (120.7 mm × 25.4 mm × 5.1 mm) specimens were tested. Two of each size were tested on a fixture with roller supports (25.4 mm diameter) to reduce friction, as suggested in ASTM E399 [18], with the remaining specimens tested on the fixed supports. All specimens were tested with a 101.6 mm span. The results are listed in Table 3.

The specimens tested on fixed supports were identical in size and tested in the same manner as those tested previously (Table 2) except that the span over which they were bent differed. It was found that the values of K_M for the medium bend specimens (108.0 mm × 20.3 mm × 5.1 mm) determined with a span = 101.6 mm fell within the scatter band of those with span = 81.3 mm. The large bend speci-

Table 3. Fracture Toughness Values Determined by Three-Point Bend Fixtures with Span = 101.6 mm and with either fixed or roller supports

Specimen L x W x T x span (mm x mm x mm x mm)	Fixture Supports	Number Tested	K_M - MPa $\sqrt{m}$ (Mean with 95% Confidence Interval*)
108.0 x 20.3 x 5.1 x 81.3	Fixed	3	54.8 ± 2.4
108.0 x 20.3 x 5.1 x 81.3	Rollers	2	53.0 ± 2.9
120.7 x 25.4 x 5.1 x 108.0	Fixed	1	66.0 ± 3.1
120.7 x 25.4 x 5.1 x 108.0	Rollers	2	57.8 ± 2.2

*Calculation of confidence interval used standard deviation estimated from the tests of all similarly sized specimens regardless of span.

mens (120.7 mm × 25.4 mm × 5.1 mm) were tested at nearly identical span lengths of 108.0 mm and 101.6 mm. The value of K_M determined from the single test at the shorter span fell only slightly (less than 1. MPA $\sqrt{m}$ or 2%) below the scatter band of K_M determined from the 6 tests at the larger span. These comparisons suggest that the value of K_M is independent of span length over the span lengths tested. Consequently, it seems reasonable to assume that the standard deviation associated with K_M is also independent of the span length at which specimens of identical size are tested. With this in mind, the confidence intervals placed upon the values of K_M in Table 3 were calculated using the standard deviation of the previously tested, identically sized bend specimens. The 95% confidence interval for the mean value of K_M as determined for medium sized (108.0 mm × 20.3 mm × 5.1 mm) specimens tested with fixed and roller supports overlap. Both test fixtures generate consistent values of K_M for this size of specimen. However, the mean value of K_M determined from the larger bend specimen (120.7 mm × 25.4 mm × 5.1 mm) appears to depend on the test fixture used, as evidenced by confidence intervals which do not overlap. It was observed that brinelling was most severe for the large specimens, and that the brinelling under the supports was greater when fixed supports were used. Greater lateral friction is present in this condition, which accounts for the higher value of K_M. The value of K_M presented in Table 2 and Figure 3 for the largest three-point bend specimen should probably be considered as representing a value with substantial brinelling. If this brinelling is reduced, the value of K_M may be decreased by 15%. This would cause an even greater difference in the mean values of K_M for center-notched and the large three-point bend specimens.

The failure of linear elastic fracture mechanics, as commonly applied to unidirectional boron/aluminum, to define a material property characterizing fracture in these specimens is felt to be due to widespread matrix yielding. This yielding occurs in regions not limited to the immediate vicinity of the notch tip. Pervasive yielding is evidenced by a highly nonlinear load versus notch opening displacement record (Fig. 1). To experimentally verify this widespread matrix yielding, a center-notched specimen was strain-gaged with 3.0 mm square, stacked rosettes on a line perpendicular to the notch tip (Fig. 4). Solid curves in the figure correspond to composite shear strain γ_c defined in a coordinate system with axes parallel and perpendicular to the fiber direction. Each curve represents the average value determined by two symmetrically positioned rosettes. The large composite shear strains indicate matrix yielding at all gage positions – even 38.1 mm above the notch tip. The existence of widespread yielding casts doubt on the utility of linear elastic fracture mechanics for this specimen, since the crack tip singularity in an elastic body has no relevance. The figure also displays a predicted relation (dashed curve) between average applied stress $\sigma_{c\ av}^*$ and composite shear strain. The form of the predicted relation compares favorably with the experimental record. These predictions are based upon a numerical model of a finite dimensioned, center-notched monolayer developed by Reedy [26]. This model incorporates the composite's heterogeneous structure

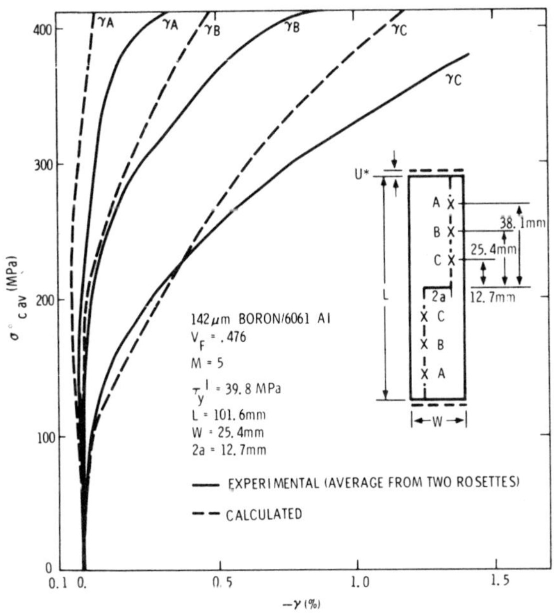

Figure 4. Composite shear strain γ at positions A, B and C vs. average applied composite stress $\sigma^*_{c\,av}$.

and permits the matrix to yield and work harden. Comparison of these results indicates that an accurate analysis of center-notched, unidirectional boron/aluminum must account for work hardening of the aluminum matrix, and permit matrix yielding to occur in regions quite distant from the notch tip. This latter consideration definitely violates one of the fundamental tenets of linear elastic fracture mechanics – that of localized plastic flow.

CONCLUSIONS

This series of tests has shown that fracture toughness as commonly defined for unidirectional boron/aluminum does not define a material property for the type and size of specimens tested. It was also found that the failure mode of notched specimens loaded symmetrically about the notch (Mode I loading) varied with specimen type. Colinear notch growth occurred in center-notched and three-point bend specimens, whereas compact tension specimens failed when a longitudinal crack initiated at the notch tip and propagated perpendicular to the notch. The inability of methods based upon linear elastic fracture mechanics to account for these effects appears to be a consequence of widespread matrix yielding. Matrix yielding in shear at regions remote from the notch tip was experimentally observed with strain gaged, center-notched tensile specimens.

ACKNOWLEDGMENT

I wish to acknowledge H. K. Street of Sandia Laboratories for the development of test fixtures and also for the careful attention with which he performed the testing.

REFERENCES

1. N. R. Adsit and W. E. Witzell, "Fracture Toughness of Aluminum-Boron Composites," proceedings of National SAMPE Technical Conference, Seattle, Washington, 1969, p. 391.
2. E. F. Olster and R. C. Jones, "Toughening Mechanisms in Continuous Filament Unidirectionally Reinforced Composites," *Composite Materials: Testing and Design* (Second Conference), ASTM STP 497, American Society for Testing and Materials, 1972, p. 189.
3. J. R. Hancock and G. D. Swanson, "Toughness of Filamentary Boron/Aluminum Composites," *Composite Materials: Testing and Design* (Second Conference), ASTM STP 497, American Society for Testing and Materials, 1972, p. 299.
4. K. G. Kreider and L. Dardi, "Fracture Toughness of Composites," *Failure Modes in Composites,* I. J. Toth Edt., The Metallurgical Society of the AIME, Vol. I, New York, 1972, p. 193.
5. F. A. Iannuzzi, *On the Fracture Toughness of Composite Materials,* Ph.D. Thesis, The University of Tennessee, 1974.
6. N. R. Adsit and J. P. Waszczak, "Fracture Mechanics Correlation of Boron/Aluminum Coupons Containing Stress Risers," *Fracture Mechanics of Composites,* ASTM STP 593, American Society for Testing and Materials, 1975, p. 163.
7. J. M. Grandemange and K. N. Street, "The Effect of Notch Root Radius and Thickness on the Static Fracture Toughness of B-Al Composites," proceedings of the International Conference on Composite Materials, Vol. 2, E. Scala, E. Anderson, I. Toth, and B. R. Noton editors, Boston, MA, 1975, p. 1019.
8. W. R. Hoover, "Crack Initiation in B-Al Composites," *J. Composite Materials,* Vol. 10 (1976), p. 106.
9. S. T. Mileiko, N. M. Sorokin, and A. M. Zirlin, "Fracture of Boron-Aluminum Composites," proceedings of the International Conference on Composite Materials, Vol. 1, E. Scala, E. Anderson, I. Toth and B. R. Noton, editors, Geneva, Switzerland, 1975, p. 562.
10. J. Awerbuch and H. T. Hahn, "Hard Object Impact Damage of Metal Matrix Composites," *J. Composite Materials,* Vol. 10 (1976), p. 231.
11. C. T. Sun and K. M. Prewo, "The Fracture Toughness of Boron Aluminum Composites," *J. Composite Materials,* Vol. 11 (1977), p. 164.
12. J. P. Waszczak, "Applicability of Linear Elastic Fracture Mechanics to 5.6-mil Boron/6061 Aluminum," *J. Aircraft,* Vol. 13 (1976), p. 770.
13. Piet Peters, "Fracture Mechanical Invesitgations on Unidirectional Boron-Aluminum and Boron-Epoxy Composites," *J. Composite Materials,* Vol. 12 (1978), p. 250.
14. J. Awerbuch and H. T. Hahn, "Crack-Tip Damage and Fracture Toughness of Boron/Aluminum Composites," *J. Composite Materials,* Vol. 13 (1979), p. 82.
15. M. A. Wright and D. Welch, "Failure of Centre Notched Specimens of 6061 Aluminum Reinforced with Unidirectional Boron Fibres," *Fibre Science and Technology,* Vol. 11 (1978), p. 447.
16. M. A. Wright, D. Welch and J. Jollay, "The Fracture of Boron Fibre-Reinforced 6061 Aluminum Alloy," *J. of Matl. Sci.,* Vol. *14* (1979), p. 1218.
17. D. Welch and M. A. Wright, "The Application of the Principles of Fracture Mechanics to a Study of the Fracture of Fiber Reinforced Aluminum," proceedings of the International

Conference on Composite Materials, B. Noton, R. Signorelli, K. Street and L. Phillips, editors, Toronto, Canada, 1978, p. 938.

18. "Standard Test Method for Plane-Strain Fracture Toughness of Metallic Materials, ASTM E399-78," *1978 Annual Book of ASTM Standards Part 10,* American Society for Testing and Materials, p. 512.
19. K. M. Prewo and K. G. Kreider, "Fatigue Failure Mechanism in Boron Aluminum," *Failure Modes in Composites,* I. J. Toth Edt., The Metallurgical Society of the AIME, Vol. I, New York, 1972, p. 395.
20. J. Awerbuch and H. T. Hahn, "K-Calibration of Unidirectional Metal Matrix Composites," *J. Composite Materials,* Vol. 12, (1978), p. 222.
21. H. Tada, *The Stress Analysis of Cracks Handbook,* Del Research Corporation, Hellertown, PA, 1973.
22. W. R. Hoover and R. E. Allred, "The Dynamic Fracture Behavior of Borsic-Al Composites," proceedings of the Dynamic Crack Propagation Symposium, G. C. Sih, Ed., Noordholf International, Leyden, England, 1973, p. 613.
23. S. E. Benzley and Z. E. Beisinger, "CHILES 2 – A Finite Element Computer Program that Calculates the Intensities of Linear Elastic Singularities in Isotropic and Orthotropic Materials," SAND 77-1900, Sandia Laboratories, Albuquerque, New Mexico, February, 1978.
24. M. E. Waddoups, J. R. Eisenmann and B. E. Kaminski, "Macroscopic Fracture Mechanics of Advanced Composite Materials," *J. Composite Materials,* Vol. 5 (1971), p. 446.
25. R. J. Nuismer and J. M. Whitney, "Uniaxial Failure of Composite Laminates Containing Stress Concentrations," *Fracture Mechanics of Composites,* ASTM STP 593, American Society of Testing and Materials, 1975, p. 117.
26. E. D. Reedy, Jr., "Analysis of Center-Notched Monolayers with Application to Boron/ Aluminum Composites," submiteed for publication.